R S Amble

Rise of the

Digital

World

An opportunity India cannot afford to miss

Gloture Books
Chicago

ISBN-13: 978-0-9831574-9-6

Published by Gloture Books, Chicago
Book cover image: Credit: AndreyPopov/iStockphoto.com

Printed in the United States of America

We owe a lot to the Indians, who taught us how to count, without which no worthwhile scientific discovery could have been made.

— Albert Einstein

Also from R.S. Amblee

The Audacity of Futurism

Winner 2017 NATIONAL INDIE EXCELLENCE
AWARDS

The Ugly Fight
*Unleashing Artificial Intelligence Against Global
Warming*

Winner 2019 NATIONAL INDIE EXCELLENCE
AWARDS

Honored Best Book Award Finalist at the American Book
Festival

Contents

Rise of the

Digital

World

Introduction
An Opportunity

Looking at the world wearing a geek's hat is an exhilarating experience. All digital enigmas will make sense if you give it a nerdy look. After completing my executive course at the Massachusetts Institute of Technology (MIT) on artificial intelligence (AI), blockchain, and the Internet of Things (IoT), I drifted into an intellectual pursuit of looking at the world from a different perspective.

Digital transformation, aka *smart transformation,* is happening all around the world and is growing. It is not just our phones that are becoming smart; homes are becoming smart. Businesses are going smart. Smart cities are sprouting up. Manufacturing and transportation are going smart too.

A smart world is rising!

The most obvious answer as to why the world is going through such a prodigious digital transformation is that consumers love the great comforts provided by smart products, and companies save tons of money by going digital. Above all, with faster computers and smart software, the pace of innovation is at an all-time high. The ease of life and the expediency they offer is unbelievable.

Just compare our smartphones with the analog phones we had just a few years ago. Just look at the convenience of online shopping, which was unheard of a few years ago. The GPS that guides us, smart thermostats that turn on the heat, so we come home to a warm house, mobile apps that let us manage our finances without going to the bank, and now the incredible online social meetings protecting us from exposure to viral pandemics—the comfort and convenience they provide us is unprecedented.

We are pulling these technologies that are already pushing upon us. Digital transformation is a no-brainer. Tech tsunamis are coming one after another. It didn't take much for the world to catch up with the smart transformation, which is now a multi-trillion-dollar industry and is permeating every industrial sector. The current digital transformation is not dependent on the industries' willingness to undergo a transformation; they have already buckled up. Its pace is now solely dependent on the availability of the information technology (IT) workforce. There will be an unprecedented demand for IT skill sets as transformation fever catches on. However, if the supply of the skilled labor force is stunted, so will the transformation.

The global transformation market is currently $1 trillion and is expected to reach $5 trillion by 2025.[1] Currently there are not enough skilled IT experts in the world to accommodate a market that size. In the United States alone, there are more than a million high-tech jobs still unfilled. This dearth of skilled IT workers has caused IT sector wages to spike, causing the cost of digital transformation to spike as well, which is killing its pace. The world needs a low-cost, skilled IT workforce for a high-paced transformation to make the world a better place for all of us. Lower wages mean businesses can hire a larger workforce. A larger workforce means only one thing: accelerated digital transformation. That is what the world is envisaging. That is where the IT transformation opportunity lies.

This book is all about these amazing smart transformations and the opportunities they bestow upon us.

However, only developing countries with a huge population, excellent IT skillsets, and lower wages can afford to offer IT skills at low prices. Wealthy developed countries simply cannot afford to offer such low wages. But India—highly populous and an IT hub of the world—could

---

[1] All currency figures in this book are in US dollars.

absolutely contribute to this digital transformation. India has the potential and vigor to provide the skills the world needs to speed up the digital transformation. India cannot afford to miss this opportunity.

In this book I share my American experiences, which to a large extent is typical for most of the developed world. My research and explorations will take you through an exhilarating journey into the digital enigma of the smart world.

Part I
Withdrawal from Globalization

Chapter 1
A Smart World Is Rising

For many people who got their homes refinanced during the COVID pandemic of 2020, there was a big surprise: most homes were appraised remotely without an inspection. COVID forced banks to take extra precautions to eliminate unnecessary tasks with the potential to expose customers and bank employees to COVID. When determining whether to refinance, banks use a loan-to-value ratio. If the value of the house is considerably high compared to the loan to be refinanced, then there is no need for a physical inspection, and the appraisal can be done remotely. For single-family homes, many times the land value itself is so high that the condition of the house becomes inconsequential.

Why weren't appraisals done this way before? COVID forced businesses to think more rationally and efficiently. However, if the loan-to-value ratio is not good enough, then they may order an inspection. But even in such a scenario, the owner of the house could give a virtual tour of the house to the remote appraiser via an app like Zoom or Skype.

The world is changing fast. For many people in-store shopping is already a thing of the past, with online shopping growing ever more popular. During 2020, online sales in the United States outperformed brick-and-mortar sales by a factor of 10. Online retail sales are expected to reach $6.5 trillion by 2023.

Grocery shopping online is another area getting a big boost. Consumers can now order their groceries online and pick them up curbside to minimize potential exposure to the coronavirus. Many customers love the comforts of home delivery even more, which can be done within a few hours of ordering. Above all they can order from multiple stores for one single delivery. Online shopping saves the hassle of dressing up and driving in traffic, especially for the elderly and sick, as well as for busy people who are pressed for time. There is a walloping demand for food delivery too. In the United States many services such as Deliveroo, JustEat, Postmates, and Uber Eats have introduced contactless delivery options to reduce the risk of spreading the virus.

These trends have no reason to revert once the pandemic ends because they save people tons of time. All these changes would have eventually happened even without COVID; however, it would have taken five to ten years or more as technology slowly evolved. COVID made it happen in one year.

During the pandemic we've been stuck at home, busy working and forgetting to drive. Car dealers have seen the effect of this, with millions of vehicles dormant across the country. They're suggesting that car owners who don't plan on driving for months should add some fuel stabilizer to ensure that their vehicles' gas doesn't go bad by the time they're ready to drive again. What a paradox we're seeing: a car country is becoming a dormant-car country.

Jobs and careers are getting a makeover too. Job interviews are done online, employees work from home, and meetings are done online—many new hires have not even seen their offices. Employers are saving tons of money by not renting office spaces. In San Francisco alone, nearly twelve-million-square-feet of office space was vacant by the end of 2020. Many employers have expressed their desire to keep this remote working option permanent as it saves them substantial amounts of time and money. Moreover, they can hire someone from anywhere in the world.

That's good news for employees too, who can work for any employer in the world while enjoying all the comforts of home. Before COVID, employees used to leave their office well before 5:00 p.m. to catch their trains or buses or to drive home. Working from home is saving employees two to three hours of commute time every day, which they are putting back into work. Recent research has shown that the productivity of companies has increased significantly as employees now work longer hours from home. There is a collateral advantage too; less travel means fewer emissions. Many cities around the globe are seeing clean air for the first time in decades.

According to some analyst groups, as of 2020 nearly 40 percent of the American workforce was working remotely. Companies like Twitter and Square are letting employees work from home permanently, and many other companies are joining this race. Microsoft, which has about 160,000 employees worldwide, is offering more flexibility to its workforce to work from home.

But the remote work culture has negatively impacted some states' economies. With employees working from home, many companies are relocating their empty headquarters to other states where taxes are lower, saving millions of dollars. For example, according to relocation specialist Joe Vranich, in 2016—the most recent year data was available—nearly 1,800 small- to medium-sized companies moved out of California—before COVID. The pace of the exodus has picked up recently as the remote work culture is making this migration seamless. Large companies like Oracle, Hewlett Packard, and Tesla made headlines recently when they moved some of their operations out of Silicon Valley.

The same is true with employee relocation. Because people can work from anywhere in the world, they can now move to the city of their choice. Unquestionably, smart cities would be on their choice list; however, among those smart cities low tax states would be more attractive. State and county administrations have to realize that the global work culture is changing, and any unreasonable tax hike could be a trigger for companies and employees alike to flee, inflicting huge financial losses to their tax base.

We are witnessing the rise of a digital world, which even without COVID was already proceeding; the virus just sped up the inevitable.

Our Own Digital Twin

A digital twin is basically a bridge between the physical and digital worlds. Our online footprint is our social digital twin. We have a virtual social presence, virtual financial presence, and we even have virtual careers engendering digital colleagues, digital bosses, and digital teams. Even the online training classes that we attend beget digital classmates, digital teachers, and digital friends we have never met in person. Every one of us has a digital twin now, which will stay online forever even after we are gone.

In the olden days genealogists had to go to libraries, historical societies, and small-town vital statistics offices to research individuals' life history. But now all that is available digitally with the click of a mouse. We are all building our digital twins. Future generations can look at our social lives, financial lives, careers, accomplishments, and so on. Artificial intelligence algorithms have evolved to such an extent that they can even build our digital voice and make us talk after we are gone. Now each one of us has truly become virtually immortal.

A majority of people crave building a digital twin in as much detail as possible. You can see the evidence of this in the number of posts on Facebook, Instagram, and other social sites. Nealy 350 million photos are uploaded every day to Facebook. Over forty billion photos and videos have been shared on Instagram since its inception. This craving to foster a digital twin is cogently helping drive the rise of the digital world.

Smart technologies have protected us immeasurably from the pandemic beyond online shopping and working remotely from home to help avoid exposure to the pandemic. Telemedicine enables us to see a doctor online. Apps like Zoom, Skype, and others help children attend online classes. And video chats keep our social life alive without physical proximity. The accessibility and services the digital world have bestowed upon us are immeasurable. Without these marvelous smart technologies, the world would have endured even greater suffering during the pandemic onslaught, and it would have taken us decades to recover.

Is it not gratifying that a smart world is rising?

The pandemic is undeniably prompting countries and cultures around the world to consider deglobalizing as protection from future viral attacks. Is this smart world ready to deglobalize? If you dig deeper, you'll be surprised at how close the smart world and the deglobalized world are.

Chapter 2
Is the Smart World Ready to Deglobalize?

Why COVID-19 May be a Major Blow to Globalization.
~ The Times magazine

A Global Outbreak Is Fueling the Backlash to Globalization.
As the coronavirus spreads around the world, companies are seeking alternatives to making goods in China,
while right-wing political parties fulminate against open borders.
~ The New York Times

During my executive course on AI and IoT at the Massachusetts Institute of Technology, the more I learned about smart technologies, the more intriguing they became. When I started writing my final assignment of putting all these technologies together, the power and dexterity of digital transformation made me look at deglobalization from a different perspective.

Early civilizations came up with the idea of globalization to prosper through trade. Nations wanted to trade their products with the best products the world had to offer. The desire to bring home something sumptuous can be traced back to the spice trade. Merchants traveled thousands of miles on horses or camels, by sea or by foot, to bring home items desired by the local populace. Today we have over fifty thousand merchant ships transporting every kind of cargo to accommodate international trade demands for the best products and services. We've come a long way to achieve this level of supremacy in global trade.

Now the world is poised to deglobalize. The COVID crisis has forced countries to close their borders and put restrictions on travel; consequently, the supply chain is breaking down. COVID made us think and act differently. Now the whole world is envisaging deglobalization to protect itself from future viral attacks and the financial disasters that follow viral events. Can just one pandemic have such a far-reaching influence to actualize the deglobalization of the whole world, or was there something already brewing, and COVID was just a tipping point?

To begin with globalization was built on unethical trade. It engendered unscrupulous issues like local job losses, human rights violations in foreign workplaces, environmental disasters like shipping container oil leaks, financial threats like currency manipulations, quality failures leading to junk products filling up our houses—you name it, we had it.

Over these past decades, we have developed a love-hate relationship with globalization. We hated it when our local jobs receded, but we loved it when we got cheap products in the market. We hated it when ships leaked oil and decimated marine life, but we loved it when oil fueled our gas-guzzling cars. We hated it when we saw low-quality products but loved it when we could not afford high-quality local doodads. We hated it when we heard of human rights violation in China but loved it when our own poor people could afford low-cost products. We loved more than we hated, so globalization survived and thrived.

Getting products from China is far cheaper than setting up local factories. It is impossible to compete with China when our own labor costs are sky-high. The only way to compete with China is to produce our own goods with less or no labor—in essence adopting full automation, which is quite expensive without the digital transformation of our industries. Unless and until our machines and robots become digitally smart, the cost of production will remain assuredly high. That is why industries around the world are now gearing toward digital transformation. For a long time we have fallen prey to cheap Chinese labor, keeping all our innovative ideas about digitization on the back burner. Now the world is repenting its stance, realizing that digital transformation is the most pragmatic way to deglobalize. However, it's not easy to come out of the globalization we've evolved so well for centuries. Shifting our industries to automation is a giant leap.

Self-reliance sounds so good, so refreshing, so humane, and ethical, but the journey is acutely challenging. Regardless, we have to discard globalization for the good of mankind. We have to dispose of it to protect human rights. We have to dump it to save our environment. There is pressure building up in the political arena, with trends like Brexit and other elections and referendum across Europe and the Americas pointing in that direction. Deglobalization in India is labeled as *self-reliance*. In the United States deglobalization is termed: *Made in America*. Every country is looking at the same concept from different viewpoints.

Even before COVID, globalization was on the verge of slowing down as wages were steadily increasing in China, and the cost of industrial robots has been dropping consistently over the last few years. The combination of these two factors, along with the US-China trade war, had rekindled the idea of onshoring, which is the practice of transferring business operations that were moved overseas back to the country it was originally located in. Even Japan has allocated a stimulus package of $2 billion to Japanese companies seeking to bring back production from China.

A Bank of America survey of three thousand global companies across twelve industries showed that nearly 80 percent of the companies were rethinking their dependency on global supply chains.

Now the million-dollar question is, would such a paradigm shift even be possible? If it is conceivable, then how does it look like in real life?

Digital transformation manifests differently depending on the industry. In manufacturing it means moving from ordinary robots to smart robots that can produce products at a lower cost and enable reshoring, providing a plethora of high-tech opportunities to local entrepreneurs that further fuels digitization. In the mining industry it means excavating raw materials using smart machines, so deglobalization happens naturally and shuns the movement of products across the border. In the energy sector it means producing power with smart solar robots, which will naturally make the fossil fuel industry obsolete.

These are not futuristic concepts; these transformations are already happening in the smart world. COVID is just adding fuel to the digital revolution fire.

Another force driving deglobalization is the discomfiting realization that global trade now has a single point of failure: China. This dependency has caused a cascading series of financial disasters, most recently after the border closures amid the pandemic. From that perspective the world needs deglobalization but cannot live without global trade; digitization gives us hope to achieve both. Digitization will not shun global trade; it will digitally enhance it. If a country wants to produce goods and sell them in the US, it can do so by producing them on US soil at a low cost with smart robots and without material movement. Countries could digitally trade with each other at an unprecedented scale through know-how and IT skillsets. That would keep healthy competition alive and innovation thriving in the world. It takes the global trade to newer heights without material movement.

In a deglobalized environment the world could come much closer to producing the best in their own yards. That's the direction Industry 4.0—the ongoing automation of traditional manufacturing and industrial practices, using modern smart technology—is heading, bringing the good of digitization without the ills of globalization.

While the rise of the smart world is helping us deglobalize, the surge of deglobalization is pushing the smart world to progress even faster. The smart world and deglobalized world are the two faces of the same coin; a smart world is a deglobalized world. COVID is accelerating the rise of the smart world, and in the same breath it's influencing deglobalization. A pandemic that took the world by storm has catapulted it decades into the future.

Digital transformation is already happening in most industries. Two of the phenomenal technologies that are driving this digital transformation are the IoT and AI. Although there is a separate chapter dedicated to these technologies at the end of this book, let's take a quick glance at these two phenomena just to familiarize ourselves as we will be referring to them in several chapters. Any digital gadget that connects to the Internet via embedded sensors—cell phones, watches, garage doors, lights, switches, TVs, refrigerators, robots, drones, tools, LiDAR, etc.—is termed an IoT. Soon anything and everything of value will be an IoT.

One challenging feature of IoT devices is that they generate vast amounts of information called big data, so vast that humans simply cannot decipher or analyze it for any actionable purpose. For instance, if you look at today's weather prediction, data is collected from various sources like space satellites, weather balloons, radar systems, and umpteen IoT sensors. Trying to analyze this kind of inordinate, ever-changing data to accurately predict the weather for extended periods is next to impossible for anyone, even meteorologists—but not for AI, which can analyze big data and provide a weather forecast that is many folds more precise than anything generated by human forecasters.

In the olden days *one-robot, one-task* used to be the norm. Programmers used to spend months—even years—coding these dumb robots. That's why robotic manufacturing was so expensive—the stark reason why the United States went to China for manufacturing. Now AI-maneuvered smart robots can learn on their own by observing other robots or humans working on specific tasks. These self-learning robots are available off the shelf. They get better and better as they learn continuously. This is a manufacturing game-changer, a fiery force influencing deglobalization by eliminating expensive programming and making manufacturing much cheaper. With smart robots, products could be produced in any corner of the country, naturally influencing deglobalization.

Opportunely, AI algorithms are being coded worldwide to process this gigantic trove of big data. Without these AI programmers, big data would not be of much value. IoT and AI complement each other: one is a data producer, the other is a consumer.

According to the "World Robotics 2020 Industrial Robots" report, there are about 2.7 million industrial robots in the world today. About 300,000 new robots are sold annually. Through AI, robots are going through a ginormous digital transformation of their own, just like any other machine.

This amazing story of digital disruption is not true in developing and underdeveloped countries, which have not undergone globalization to the extent developed countries have. Their economies, not churned up by globalization, are still slow, and their industries are still laid back because many don't have exposure to AI and robotics yet. The only way for these countries to achieve self-reliance is to skip the globalization that developed countries previously experienced and jump straight into digitization.

India has done this before; the country skipped the era of personal computers and jumped straight into the meadow of mobile phones. Now it's attempting yet another stunning jump. It doesn't need to bring in Chinese products to impel its economy. India has the potential to digitize its industries and the world at the same time.

For the world to deglobalize, it needs many, many skilled IT professionals, which are in demand and hard to find. And even if they are available, they are quite expensive, especially in developed countries. That is why digital transformation costs are quite high and why digital transformation is progressing at an exceptionally low pace.

Could India, the IT hub of the world, engender the skills that the world needs, and could deglobalization be an opportunity for India?

Chapter 3
India's Deglobalization Opportunity

If we had talked about deglobalization a few years ago, it would have meant insulation from the rest of the world. Today it means accelerated globalization via digitization, which empowers countries to produce products at very low costs within their own borders, eliminating the movement of materials between countries and protecting them from future viral pandemics. Both developed and developing countries are ready for such a sea change. But digital transformation needs a large IT workforce that is currently unavailable. It's a void India is uniquely positioned to fill based on three realities.

First and foremost is the impressive size of India's existing IT workforce, with more than five million professionals employed in about eighty countries around the world. There are more than one thousand global delivery centers across the world, run by two hundred or so Indian IT companies. IT firms including Infosys, Wipro, TCS, Tech Mahindra, and Cognizant are already offering artificial intelligence, blockchain, IoTs, cloud computing, and other services to global clients. But while five million Indian IT professionals may seem prodigious, it's a fraction of the global demand out there.

Secondly, India is advantageously positioned with its higher education institutions, which produce more than 350,000 engineering graduates every year. Of India's eleven thousand colleges, 1,200 are for engineering. Not only is India already at the forefront of global IT education, but it's marching to become the educational hub for the world.

The third advantage is India's huge young population. India is the second-largest populated country in the world, only behind China. According to the 2019 census, about one-third of India's 1.36 billion people are below fifteen years of age. Four hundred million is a colossal youth population. Most reside in poor villages, but poverty in India has a different meaning than anywhere else in the world. The hunger for food is rapidly morphing into a hunger for education and careers. Village youths by the millions are migrating to booming urban areas. These young people have realized that education is the best way to pursue a prosperous career and are going in search of green pastures in cities far away from their rural villages. Even poor parents are moving to cities, hoping to provide their children with better educational opportunities. India is the only country where the hunger for food is an ardent driving force to learn skills to forge a career. To complement this burning passion, a plethora of educational institutions are springing up.

It is a rare but perfect storm: a highly energetic IT force ready to build the digital world, poor and hungry youths ready to learn any skill, and an abundance of intensely motivated schools and colleges. This vibrant template is what India can offer to the world.

As previously mentioned, experts predict that the IT service sector for the digital transformation market could hit $5 trillion by 2025, which is more than 8 percent of the total GDP of all the countries on Earth. Considering the average annual salary in India is equivalent to about $5,000, a $5 trillion market could accommodate nearly one billion IT workers in the subcontinent. However, in the United States and other developed countries, where the average salary is about $100,000 a year, there is the potential for only fifty million IT employees to carry the projected market size. In other words, high wages mean a stunted digital transformation. Imagine the pace of global digital transformation should India fill the void with its low-cost IT skills.

If four-hundred-million Indian youths are trained over the next decade in the IT field, each earning $5,000, it would collectively add $2 trillion dollars to the Indian economy. This revenue is apart from the corporate profit that Indian companies would make. Together that would add trillions of dollars to the nation's economy just by educating village youths. Imagine the impact of this revenue on the nation's economy as it exponentially increases the living standards of the poor.

There are more than six-hundred-thousand villages in India, mostly poor, which hold 65 percent of the youth population. But in the digital transformation, the dusty, poor, backward villages of India will not be so poor or so backward after all because their earning potential is enormous, and they will shine as hidden gems. In a typical village of one hundred youths, the collective windfall could exceed half a million dollars in one year. Then consider the spending capacity of this group; like workers anywhere else, they would buy goods to increase their standard of living, which in turn would have a profound impact on the nation's economy. Like ripples from a stone thrown in a lake, it all emanates from the digital transformation.

Though this possibility is so exhilarating to hear, it would not come without consequences, both good and bad. There will be three major ramifications for the Indian economy as it goes through the revamping of its village education. The first and foremost is a demographic shift. After receiving sought-after IT skills, migration away from villages to urban areas is an inevitable part of the digital transformation. Having millions of youths relocate to cities will create significant labor shortages in agriculture and will lead to the eventual digitization of agriculture itself. It will also force cities to go digitally smart to manage the giant influx of people, further opening digital opportunities.

The second impact is on India's own industrial digitization. India's manufacturing sector has the potential to reach $1 trillion by 2025. When an abundant, low-cost IT workforce becomes available, Indian industries will be motivated to embrace a digital strategy to achieve self-reliance. Not many industries in India are currently digitized; instead, they're still using old technologies. Mega industries like iron and steel, textiles, jute, sugar, cement, and paper would go through a massive digital transformation. Other giant industries like petrochemical, automobiles, and finance have begun the process of digitization to some extent but have a long way to go. When those giants go through digital innovation, there will be a plethora of job opportunities for IT professionals.

The third impact will come from start-ups. In 2019 nearly 1,300 new technology start-ups sprang up; that works out to two or three tech start-ups established almost every day. Tech start-ups have an advantage in that they achieve self-reliance quickly, challenging the status quo. Many long-established industries in India have not started any digital transformation, so the process of mixing both worlds could be a little chaotic.

Village education in India is going to have an unprecedented impact on the global economy. Many countries will seek out Indian IT workers as self-reliance catches on. From that viewpoint India has a dual role to play. While Indian IT specialists digitize industries around the world, Indian industries would also go through equally massive digitization. That is where the aspirations of the world and the aspirations of India will converge. In that context the more automation the world experiences, the more automation India will experience, so its self-reliance irrevocably intertwined with the world's.

For India this is an opportunity, a responsibility, and a technological challenge all bundled together. It's also an ethical obligation to fill the IT void for a greater cause, which is why this is the right time to fortify education for village youths: it promotes self-reliance, and a self-reliant world is a safer place to live.

Even without the pandemic, the world would have eventually become digitally smart and deglobalized anyway. However, what we are witnessing today is a conscious effort from a COVID-wreaked world to become digitally smart sooner. The upcoming chapters explore how smart the world is right now, where the smart transformations are happening, and how they all fit into this deglobalization quagmire. Through my research and exploration, the book offers a glimpse into potential opportunities beyond India to the entire global tech diaspora, a journey across the smart world.

Part II
The Dawn of SMART Dogma

Chapter 4
The Rise of Smart Cities

As of early 2021 there were about 102 smart cities in the world, but their numbers are growing. The world's top six hundred cities, which command about 60 percent of the global GDP, are hammering away to become smart, defined as solving problems through technology.

The Switzerland-based International Institute for Management Development, in partnership with Singapore University of Technology and Design, compiled the "Smart City Index 2020."

When a city is designated smart, that fundamentally means it's a safer, comfortable, and upbeat place to live. It unequivocally guarantees cleaner air and water. It emphatically offers efficient utilization of public facilities, including healthcare. The digital smartness of a city significantly boosts the quality of life.

A smart city is a collaborative effort between the officials who run the city and the people who live in it to make a city smart. Both should produce useful data from a variety of sources and share it. This includes data from individuals, homes, schools, hospitals, offices, supermarkets, retail stores, restaurants, vehicles, traffic signals, utilities, power plants, factories, and a host of other facilities. All this data is used to monitor everything from traffic to crime. To build a smart city, all stakeholders—from municipal administration to residents, from business owners to homeowners—must adopt and be adept at learning new technologies.

IoTs are the key to this smart city conversion. For example, IoTs are installed in public buses and trains so commuters can track them. That will reduce the amount of time spent waiting. So commuters and others who use mass transit can plan their day more efficiently. Smart cameras in schools are great examples of IoTs that can detect a disturbance such as an active shooter or vandalism and alert law enforcement. Smart technologies are unquestionably making schools safer. Many cities are using AI to detect crime before it happens by using the real-time data provided by security cameras and knotting it with data from social media sites.

Almost anything can be smart. Fire extinguishers in buildings are transforming into smart devices with AI and IoT sensors. The AI is trained not only to detect fire when it happens but also to identify false alarms and prevent people from being woken up in the middle of the night. AI can also predict potential fire dangers by using sensors that detect dangerous gas leaks and notify the appropriate fire, ambulance, and police authorities even before a fire breaks out.

Energy companies are installing residential smart energy meters that track the energy needs of each house to know the power demand of the entire city in real-time. This data can be used to avert power outages on days of heavy usage, such as during a heat wave in the summer, by adjusting the smart thermostats of smart homes to lower the loads. But this is only possible if the city has enough smart homes.

Street parking is becoming smart too. In many larger cities, particularly in downtown areas, it can be hard to find parking spaces. People drive around in circles looking for parking, creating congestion as well as frayed nerves. Cities are now providing smart parking spots so that drivers using their mobile app can more quickly find parking, thereby reducing the congestion. This also avoids unnecessary car emissions as well as reducing accidents and people's stress levels.

Another great smart city feature worth mentioning is air quality analysis done with smart sensors that are placed in strategic locations that detect and record contaminant particles to determine pollution levels. AI can even proactively send alerts to the facilities that are causing environmental damage so corrective actions can be promptly taken, saving the regulatory authorities the time and money of doing investigations.

Air quality sensor IoTs can also detect gases coming from nearby forest fires, so authorities are alerted well in advance, which can save lives and property.

You may have heard of old bridges becoming weak and collapsing, causing death and destruction, particularly after an earthquake or a severe flood. These kinds of unfortunate scenarios could be easily avoided by constructing bridges using smart steel and smart cement embedded with IoT sensors that could constantly send out data about the strength of the bridge. If a bridge is weakened or compromised, it would be reflected in the data, and remedial actions could be taken.

Smart cities can also respond quickly to natural disasters like hurricanes, earthquakes, floods, and even pandemics like COVID by exchanging data on resource availability such as ambulances, hospital beds, and healthcare personal, especially during nights and weekends when hospital staffs are spread thin.

The bottom line is a smart city is not just about having great facilities; it's about exchanging data among those facilities so that citizens are well taken care of. During normal days a smart city is all about saving energy and making life comfortable by efficiently utilizing resources. During emergencies a smart city is all about exchanging data instantly to quickly facilitate needed services.

Because of the lifestyle and security that a smart city offers, the real estate value in smart cities keeps going up, making the cities richer. It's not just safer and healthier to live in a smart city, it's digital fun as well. Everyone wants a piece of this smart pie, which is why every city in the world is striving hard to become smart.

What It Takes for a City to Go Smart

Smart city transformation is quite expensive, which is the main reason why bigger cities that have more cash on hand are going smart much earlier than smaller cities. Also, bigger cities tend to attract more population migration, and their resources are more challenged. On the other hand many smaller cities are experiencing political and bureaucratic backlash over the idea of going smart, elected officials shying away from the high upfront costs.

The good news is that the cost of smart devices is coming down as hardware prices are plummeting due to smart manufacturing. This is helping many cities to go smart at a lower cost. However, smart city conversion is not easy. To make a city smart, you need myriad IoT devices to be installed, configured, and interconnected with data extracted and processed. An IoT sensor doesn't do anything magical; it only generates data. Its real value is only realized when AI is adopted to handle big data. In essence a smart city transformation heavily depends on an IT skill force.

There are also risks in building a smart city. If the smart devices fail for any reason, there will be more collateral damage than otherwise. Once a city becomes smart, it becomes susceptible to cyber-attacks, which, if successful, could have disastrous consequences. We have seen many such attacks resulting in power and communication network failures.

Going smart is not just a one-time event. It needs constant maintenance and upgrades. It needs a highly skilled workforce to manage the high-tech devices. It is a serious commitment, and there is no going back.

The good news is as more cities become smart, they'll serve as a template for other cities to estimate and assess their return on investment. If a city wishes to go smart, it can choose from a host of options from other smart cities that best suit its needs. The other good news is, developing smart facilities is repetitive work as they would be implementing the same smart technologies in multiple cities, reducing overall costs. So as the smart city craze catches on, every city's tale helps every other city. Sharing information will help the smart city revolution to accelerate, making our lives safer, healthier, and more joyous. This trend is expected to engender a new breed of smart governments that are more agile and nimble.

In developing countries like India, massive migration from villages to cities is happening at an unprecedented rate, and the city resources are getting exhausted at a much faster pace. City administrations are struggling to cater to the increasing population, which requires expanding highways, constructing new housing, more hospitals, new schools and colleges, and so on as public facilities are in greater demand. If you look at how Indian cities are grappling to accommodate this growing population, the smart city concept is a boon in disguise.

In this era of climate change, saving the environment is attracting more attention than ever before. As smart cities use less water and energy and decrease levels of pollution, they will unquestionably contribute toward a healthier environment. This is one of the top factors driving elected officials of global cities to favor smart city investment. And as smart cities catch on, the world is bound to become a safer, healthier, and more joyous place to live.

Within this evolution is where India has an epic opportunity that it can't afford to miss.

Chapter 5
The Rise of Cozy Smart Homes

Smart homes are indeed cozy. Their features are almost a prerequisite for top-tier luxury real estate. As smart homes save tons of money by consuming less energy, from an economic point of view, a smart home is a smart investment.

With this kind of selling proposition, many companies are coming up with zillions of smart home gadgets: smart thermostats, smart lights, smart switches, smart refrigerators, smart energy meters, smart water meters, smart gas meters, smart laundry machines, smart dish-washing machines, smart garage door openers, smart security cameras, smart locks, smart smoke detectors, smart TVs, smart entertainment systems, and a host of others. Pretty much all electronic gadgets in our households are becoming smart. While some of them save energy, some save time. The environmental benefit of a smart home is an added bonus.

These myriad smart devices throw gargantuan amounts of big data. Artificial intelligence uses this data to understand our habits and provide proactive services. Smart thermostats can be controlled and programmed from mobile phones from literally any place in the world. Many households have gotten used to giving voice commands to smart devices like Alexa, Siri, or Google Home to change the thermostat settings. Sometimes this kind of tweaking is harmful. If tweaked too frequently, it may cause a spike in energy usage rather than savings. HVAC equipment may even break down if the thermostat is overused. This is where AI becomes useful. AI can quickly learn what time we arrive home and adjust the thermostat accordingly. These are called learning thermostats. If everyone in the house leaves, motion sensors signal the AI to adjust the temperature accordingly. Even small tweaks can save money. A single degree can reduce power consumption by 2 to 3 percent.

Similarly, smart lights also save tons of money. Motion sensors can detect whether or not a room is occupied and adjust the lights accordingly. So while you're on vacation, you don't have to worry about gadgets spiking your energy bill. At the same time if there's a burglary attempt, smart cameras will capture the crime and alert authorities.

Smart refrigerators are another great way to save time and energy. The IoTs not only regulate the inside temperature but also monitor the food you have in the refrigerator. When you run low, you can order groceries on a phone app. For this kind of automated grocery detection service, specific groceries have to be kept in specific trays in the refrigerator so that IoT sensors in those trays identify them by measuring their weight. So eggs need to be kept in egg trays, milk needs to be kept in milk trays, and so on. Instead of opening the refrigerator door to check how many eggs are left or how much milk is left, you can look at them on your smartphone from anywhere in the world.

Smart energy meters are becoming common in most cities. The data from these meters not only help balance electric loads to reduce blackouts, but the data is also sent to consumers so they can adjust their habits to lower electricity bills. Remember, the cost of electricity is high during peak times. Simply making little changes can save tons of money, such as doing laundry or running the dishwasher during off-peak hours when energy demand is lower. You can program your smart lawn sprinkler to run during off-peak times as well.

Smart water meters are now in every home. They can monitor the pressure in the system and help identify leaks in pipes and distribution lines, saving tons of money. These meters are so precise that they can identify even the smallest leaks and alert the homeowner or management company, catching the problem early to avoid a major breakdown causing expensive damages. Smart meters also eliminate expensive manual reading along with inaccurate billing and related financial losses.

Smart home entertainment systems are proliferating more than ever before. A multi-room smart entertainment system allows the user to control all connected devices so they can be listened to or watched from any room in the house simultaneously or independently. This has also helped the market for smart plugs and smart hubs to grow.

Smart garage door openers, smart security cameras, and smart cars provide both security and convenience.

Solar panels are becoming synonymous with smart homes. In cities where there is abundant sunlight, the return on investment for solar arrays is high. Smart meters are especially useful when a home has a solar array because any excess energy produced can be sold back to the local utility's grid. If more homes in a city have solar, there will be a reduced load on the grid, which could eliminate the need for—and capital expense of—building new power plants.

The other great advantage of solar is that when there is a power outage, solar homes will continue to function because they've collected and stored enough solar power in batteries. Without any external energy supply, these homes can function and keep people comfortable for hours or days, depending on the battery capacity. In developing countries like India, where there are frequent power outages, solar is an especially economically viable alternative to gasoline or diesel generators.

The global smart home market is about $70 billion and is expected to grow to more than $300 billion by 2025. As the demand for smart devices is increasing, manufacturers are finding the incentive to produce even more smart devices.

The growing work-from-home culture is putting more focus on home comforts, which is helping drive the smart-home market, creating an unprecedented demand for skilled personnel to design, manufacture, and monitor these smart devices.

The current smart home market is led by Siemens, Honeywell, Johnson Controls, United Technologies, Apple, Google, Amazon, Bosch, ABB, Ingersoll-Rand, Samsung, GE, LG, and a host of other international players. All companies are now adopting strategies to produce smart products, and company mergers, acquisitions, partnerships, and collaborations are all growing around the smart market.

The emerging smart home market is creating a tremendous need for a skilled IT workforce for mobile apps, networking, security, data analytics, and a host of other data-science expertise along with IoT competence are needed to serve this market. The demand for an IT workforce would be phenomenal—and an opportunity India simply cannot afford to miss.

Chapter 6
The Rise of Smart Businesses

Businesses are becoming digitally smart to boost their profit margin significantly higher and stay competitive in the market. If a company is not smart, inherent inefficiency will eventually drive it out of business.

Let me give you a real-world example that shows how smart technologies could turn an ailing business into a profitable one. Say a farmer is growing spinach and selling it to outlets around the country. Although the production is steady, profits are always inconsistent because there is a multitude of parameters that impact profit every year, such as extreme climate conditions. This is especially true now due to the effect of climate change. Say the farmer keeps incurring losses whenever there is extreme heat in summer or extreme cold in winter. The delicate spinach leaves are spoiling either in warehouses or while in transit. The farmer has no idea what trucks or warehouses in the country are having problems maintaining proper indoor temperature and humidity. Complaints that come from retail outlets about products after they're delivered come too late in the game because it's too expensive to replace them. That poses a challenge to the very survival of the farmer's business.

This is where digital technology will give the business the edge to survive.

The digital scenario in this case would be as simple as introducing IoT sensors into the product packaging so that temperature and humidity data for each and every produce box will be monitored in real-time. That information will help the farmer avoid any warehouses or trucks that don't maintain proper shipping conditions.

Furthermore, if the IoT data is captured and stored in a cloud-based blockchain, the farmer can even set up to pay off those businesses instantly without any paperwork and manual intervention. Further still, autonomous seeding and smart harvesting machines, which are now common, enable the farmer to manage the entire business with high efficiency from a laptop or a mobile phone. The farmer can now fully focus on business strategies rather than being stuck with inefficient management and financial losses.

Many times even very successful businesses suddenly fail, and the owner has no idea what went wrong. They need to be proactive. Businesses shouldn't depend on customer feedback because many times customers don't complain; they just move on to another business. This is where customer relationship management (CRM) software gives businesses that edge to succeed by sniffing through social media sites, discussion forums, and so on to get that vital feedback—be it compliments or complaints—and give an opportunity for businesses to improve.

Data from social media also help businesses identify the potential demographic areas for target marketing. Social CRM software such as Salesforce, Hubspot, Zoho, Sprout, Nimble, and others are now heavily used in most industries to stay competitive.

The other area that smart businesses focus on is mail processing. Companies receive tons of queries about their products and services, and if they don't respond in time, they could lose opportunities. AI-driven smart mail processing software can read even human emotions in emails by scanning for specific words. Emails with complaints are diverted to support departments. Queries are sent to the marketing department. If there are any billing questions, they are sent to the accounting department. AI software can even track customer responses, schedule follow-up meetings, and send automated reminders so that customers are efficiently serviced.

Such software saves valuable time and makes businesses efficient. A small company of ten employees could be functioning as a company of a hundred employees if they are empowered with such cutting-edge software. An ailing business could be turned around quickly into a robust, reliable, and easily manageable enterprise if it undergoes digital transformation.

In today's world all businesses are going through digitization to whatever extent technology allows them to. I've given only a few examples here, but in the real world there are hundreds of areas in businesses that have the potential to go smart. There is intense competition in digital transformation because smart software is now available at very low prices in the cloud. Businesses don't have to buy them; they can just rent them on the cloud and adopt them into their business strategies.

The effect of this digital transformation is that employees are under pressure to learn new skills. Career development is now geared more toward skills in digital transformation. Future entrepreneurship is not just about setting up businesses; it's to set them up "smartly" to survive the competition.

Smart businesses are creating a tremendous demand for a skilled IT workforce. The global demand for an IT workforce would be unprecedented—and an opportunity India simply cannot afford to miss.

Chapter 7
The Rise of Smart Manufacturing

The next revolution in industry will be data-driven smart manufacturing. Smart manufacturing uses IoTs to breathe life into machinery. For example, in a welding operation where robots weld with great precision, introducing IoTs into the robotic arms and other elements in the process will produce data that enables AI to improve all aspects of the operation: welding angle, welding temperatures, voltages, and so on without any human intervention. No matter how precise the non-smart robots are, they cannot learn the way a smart robot can.

Smart robots also leave their own digital signature. In the above example, for instance, if a defect is found during a quality assurance inspection, it's easy to identify the smart robot that did the welding and what parameters were used so it can be digitally corrected. As metallurgy advancements are made, AI can suggest better welding materials. Compare this to non-smart processes, which are hard to improve.

On a factory floor if a critical machine fails, it will have serious financial consequences. Despite routine overhauls, there are always surprises, and many robots and machines fail during regular operations. This is where digital transformation can reduce the losses. If vital parts of robots and machines were planted with IoT sensors, AI could predict such failures well in advance so that corrective actions could be taken during routine maintenance with less impact on the production. Plus, as all IoTs interact with each other, the real-time data would flow between processes, making the whole manufacturing system come alive. Essentially the IoTs and AI are breathing life into dead processes.

In today's world every industry is inevitably moving toward real-time data capture using IoTs. A typical chemical plant has a vast number of valves, gauges, switches, and so on. A typical textile plant has umpteen spinning, combing, winding, roving, and weaving operations. A typical manufacturing plant has myriad processes like turning, milling, drilling, grinding, forging, broaching, sawing, painting, laser cutting, and so on. By incorporating IoTs into devices, AI can efficiently manage processes with minimal human intervention resulting in higher quality and lower cost. Digital transformation is happening in all industrial sectors around the world to improve profit margin.

The digital twin concept is becoming immensely popular in the manufacturing industry as well. In manufacturing, a digital twin is nothing but a virtual model of a machine or a process. So if a manufacturing system were to fail, engineers don't have to open the machines and investigate the hard way. Instead, they can examine the digital twin of the same machine on their computer, figure out where the problem is, and fix it digitally. If that works, then they would fix the physical machine on the floor. This digital twin approach is extremely efficient, saving a lot of time and money.

Another exhilarating advance in the manufacturing world is the introduction of Microsoft's HoloLens, a mixed reality (MR) concept where real-world and digital objects interact. This device is designed as wearable headgear, so suppose there's a defect in an outmoded instrumentation panel that the operator doesn't know how to fix. When wearing the HoloLens a screen pops up that superimposes the instrumentation panel and identifying all the wires in the panel. Using these virtual screens helps operators easily fix the issues.

HoloLens can even create a virtual twin of a machine. By using finger and hand movements, users can touch, grab, and move the parts in the virtual twin or holograms as if they were responding to real objects. Imagine being able to pull the entire engine out of a car and turn it around to inspect it in an augmented reality experience. Imagine being able to slide your hand to modify it in real-time. That is today's smart machine design.

In industries still using a traditional non-smart method, engineers design the machine, manufacture the parts, assemble them, and then test their functionality. If there's a problem, they go back to the drawing board. In the digital twin approach, machine design is done virtually, parts are assembled virtually, and testing is done virtually. If there are any issues, then the design is fixed virtually. The physical parts are only manufactured after the digital twin is successfully tested. This approach is changing the landscape of the manufacturing industry.

The 5G or high-frequency network is adding fire to this digital transformation as its high-speed data helps machines interact with each other much more seamlessly.

By 2025 the global market for industrial robots is expected to reach about $75 billion, IoT in manufacturing $50 billion, and virtual reality in the manufacturing industry $14 billion. As the numbers show, businesses around the world realize their very survival depends on smart technologies and the IT workers who will run them. The abundance of opportunities for IT hubs like India cannot be overstated.

Chapter 8
The Rise of Smart Logistics

When you order a product online, it gets delivered in a couple of days. The supply chain that gets it from the factory or vendor to your doorstep can be simple or complex. Sometimes, although rare, a truck might bring it directly from the manufacturer; that is more likely with a start-up. But with most businesses it's more likely it goes through one or more sorting warehouses or distribution centers before landing on your doorstep. There are around twenty thousand warehouses and fifteen million trucks in the United States. This world of logistics where products in the supply chain are moved, keeping the cost to a minimum, can look like a jigsaw puzzle, which is why it heavily depends on accurate data.

A new element was recently added to this complexity: climate change. Floods, hurricanes, landslides, forest fires, blizzards, and other extreme weather can cause severe disruptions to logistics. This makes it more critical to get the best real-time data on weather, traffic, road conditions, and so on. Ultimately it is the accuracy of data that keeps businesses competitive.

For years transportation companies managed supply chains using logistic software, but it had limitations. So during natural disasters trucking and logistic companies lose money because of delayed deliveries and supply chain bottlenecks. And their insurance premiums go through the roof. Digital transformation can make transportation and delivery run much more efficiently through three smart technologies: IoT sensors, AI, and radio-frequency identification (RFID) tags which are updatable bar codes. Unlike IoTs, RFIDs are not connected to the Internet but can be read by an RFID reader. Most trucks are fitted with RFID tags to track them when they go in and out of warehouses and retail stores. Varieties of IoTs are fitted onto trucks that can collect a variety of data like the indoor temperature of the truck, outdoor weather conditions, truck speed, routes taken, warehouses visited, products delivered, products uploaded, and so on.

As there are millions of trucks on the road, the cumulative data collected from them is humongous, so implementing AI into logistics is a natural and inevitable evolution that will make supply chains more efficient, which in turn will keep costs down. For example AI will analyze weather forecasts and route trucks so they can make the best time, avoid weather delays, and keep losses to a minimum. Real-time data coming off IoTs and RFIDs give customers the ability to track their orders. Truck speed and breaking data help drivers improve their driving habits, which could reduce insurance and hospitalization costs, further reducing the insurance premiums.

If AI detects that a refrigerated truck's indoor temperature is rising, endangering its perishable cargo, trucks can be rerouted to a nearby temperature-controlled warehouse where the products can be safely stored while the truck is repaired or a replacement truck is sent for.

Warehouses are also going through digital transformation. Robots are being employed for automatic order fulfillment and moving products from shelf to shelf. Every warehouse is its own complex logistics world.

The demand for products keeps changing from season to season. Logistics also change as the demand for products changes. This is where the predictive analytics feature of AI helps logistics by forecasting the market demand.

3D printing is yet another technology that is changing the landscape of logistics. Why ship products around the world if they could be 3D printed at the destination itself? Currently 3D printers can only print certain sizes and with specific materials; however, the future is indeed bright for this technology.

The other technology that has a bright future in logistics is autonomous vehicles. The American Trucking Association estimates a shortage of almost two hundred thousand truck drivers in the United States by 2024, meaning self-driving trucks could become a necessity for the trucking industry to survive. That makes the supply chain look more like a robot world. Autonomous vehicles in warehouses combined with delivery bots and drones would make the logistic industry fully autonomous in the near future.

Supply Chain Size

The current global logistics industry is generating somewhere around $10 trillion annually. In the United States alone, the logistics market is around $2 trillion. The American economy depends heavily on trucks—and the three and a half million truckers driving them—to move most of the freight in the continental US.

The digital transformation of the transportation industry demands skills in IoT, AI, data science, robotics, cloud computing, and other cutting-edge technologies. This is a great opportunity for India to offer the needed IT skills to help digitize the global logistics industry.

Chapter 9
The Rise of Smart Healthcare

The COVID pandemic has catapulted healthcare at least a decade into the future. Healthcare would have reached this level of dexterity on its own at some point in the future; however, COVID made it happen in just one year. The global smart healthcare market is expected to reach USD $500 billion by 2025.

There are three areas in healthcare that were propelled by COVID: smart hospitals, remote patient monitoring, and telemedicine.

During the initial days of the COVID-19 pandemic, when the number of patients with severe illness grew, hospitals had to manage dwindling resources, including beds, medications, respirators, ventilators, and protective gear. Doctors and nurses had to work extra shifts over the course of months. Hospital administrators turned parking lots into testing zones and emergency rooms into makeshift ICUs.

The COVID-19 pandemic accelerated innovation at hospitals across the country by forcing them to adopt digitally focused ways of delivering care. This was accomplished by a massive data capture of treatments, recoveries, and deaths, which was shared with various city, state, and federal agencies attempting to make sense of the virus world. Remote patient data became vital to prioritize who should be admitted to the ICU and who could wait. Patients were dying by the thousands each day across the nation. The digital data of each case was crucial.

All this put stress on existing systems, forcing many hospitals and care facilities to upgrade or implement new digital systems. The lessons learned through the COVID emergency and the digital systems took the healthcare system far into the future. Even after COVID is gone, the face of healthcare will not be the same because it's evolved into a truly smart system.

The second impactful change was remote patient monitoring (RPM). Thanks to the RPM technology that had just surfaced, it was being adopted in many hospitals as a convenience to the patients so they could be monitored from the comforts of their homes. During COVID times, this new technology became a survival tool for the hospital staff who have been flooded with COVID positive patients. Patients were monitored remotely in the hospital and outside as well after they were discharged. RPM devices were also used by ambulance services so that patients could be monitored on their way to the hospital. The global remote patient monitoring market is projected to reach $100 billion by 2025.

The third area impacted was telemedicine or telehealth. During COVID times no one wanted to go to hospitals or a doctor's office for non-emergency illnesses. For minor symptoms such as headaches and colds, most people favor telemedicine, where they confer with a doctor either on a computer online or via a mobile phone app. After the consultation, a doctor can prescribe medication, which can then be delivered to the patient. Although this technology was there before COVID, it has become much more accepted and used since the pandemic. As video conferencing became a way of life, health networks such as Kaiser Permanente, Aetna, and others encouraged physicians to contact patients via phone or video calls instead of in-person clinic visits. This has drastically reduced the number of hospital visits, reducing exposure risks. The global market for telemedicine is anticipated to reach $200 billion by 2025.

While RPM devices collect patient data and telemedicine establishes the virtual patient-doctor relationship, there is a tremendous potential for AI to process this mammoth big data and become involved in remote patient diagnoses, taking healthcare to a higher level.

It is worth mentioning the effect of COVID on preventive medicine, a relatively new approach to primary care medicine, which began as a way to curb the ever-growing cost of healthcare due to chronic diseases. Although preventive medicine was encouraged, there wasn't much growth in that area as it depended heavily on patient data capture in real-time using RPM devices. Now in the wake of RPM and telehealth, preventive medicine will get a boost resulting in a healthier population.

It might have taken decades to reach this level of digitization in healthcare, but the necessities created by COVID made it possible in just one year. Investing in digitization to develop a smart healthcare system will complement digital advances, especially in areas like robotic surgery, nanotechnology, and so on.

In this sea of digital transformations, it's apparent that healthcare smartness is heavily dependent on IT skills in hardware, software, data capture, and data analytics, which are all done remotely by health IT professionals. Through focusing on these needed IT skills, India can provide its youth with an opportunity to participate in the digitization of healthcare systems throughout the world.

Chapter 10
The Rise of Smart Solar

Many news outlets have reported on the goal of completely replacing fossil fuels by 2035. If credible, that is great news. As of today the total power capacity of the world is about 4,000GW, and if we include the crude oil that ends up in our automobiles, the total power capacity averages to about 6,000GW. Renewables make up about 30 percent of the global capacity, which means we still have 4,000GW of fossil power yet to be replaced. Plus, global power is growing at about 120GW per year, which means if the global power capacity is 6000GW this year, it will be 6120GW next year, 6240 the year after, and so on. So the renewable capacity should grow beyond 120GW per year as well to replace the existing power plants.

Over the past couple of years, there were rays of hope that the 2035 goal could be achievable. For the first time solar and wind together exceeded 120 GW, surpassing that hard-to-beat number. This shuttered some coal plants. Globally, 20GW fossil plants were shut down. Out of this, 8GW of the closures were in Europe and 5GWs in the United States.

China, however, continues to dominate coal power development at the same time it is building solar power plants, but not significantly enough. Nonetheless, the solar industry is growing in China. India, on the other hand, has no new coal constructions to its credit. About 1.2GWs were closed down, and more than 27GWs of proposals were canceled.

All these developments around the world are giving hope that 2035 could be a possibility. However, in the coming post-COVID years, if there is an economic boom, which is most likely to happen, and if solar and wind don't scale up quickly, then coal will invariably make a comeback. Then the 2035 target will move further out.

Existing coal power plants will only be decommissioned after they age out, which is typically about twenty years. Once built, it becomes both an economic and a political challenge to shut them down prematurely.

You may argue that hydel (hydroelectric) and nuclear could add muscle to the renewable revolution. The truth is they have their own hurdles. Amid climate change hydel is facing either dried-up rivers or heavy flooding that threaten the safety of dams. And the safety of nuclear is still being debated. It's not surprising that the mantle has fallen on solar and wind energies.

One of the fascinating things that is happening in the solar industry is the arrival of smart robots, particularly in repetitive tasks like frequent cleaning of solar modules and continuous modulation of solar panels to always face the sun to capture more sunlight. It's great news that companies like Alion Energy, DPT, ONergy, Chemito, OMC Power, Korte, Zeton, SMA, and others are using robotic technology to some degree.

It's been estimated that with the same number of employees, the installation capacity of a solar company increases by seven or eight times if robotics is adopted. That means the current growth of the solar industry, which is struggling at 120GW, can be quickly scaled up to 840GW or more per year just by adopting robots. Despite such an advantage, the robots have failed to speed up the growth of the solar industry as programming robots manually remains expensive. Moreover, adopting robotics in the solar industry is not an easy task. Installation processes are complex as each site is unique. Robots must be reprogrammed for each site. It's not surprising that most solar installations are still manual, and because of that the solar industry may lose its edge to scale up quickly in the event of a global surge in power demand.

This is where digital transformation comes in as a savior of the power industry. Solar will go through a paradigm shift if smart, self-learning robots are adopted. Although smart techniques are still being explored in the solar industry, digital transformation unquestionably offers us the hope to replace fossil fuels. Digital transformation is being experimented with within the solar industry by using drones to collect location data from both rooftop or open land installations, for site orientation, and a host of other topographical details. Drone videos and photographs are then used to create AI algorithms to maneuver smart robots. Human-robot teams are then trained on simulators. While robots would handle mundane tasks, humans could focus on complex installation scenarios.

Although these digital transformations are still in exploratory stages, AI-trained robots and other smart approaches have amazing potential to become a great strategy to scale up the solar industry and make 2035 a reachable target—as long as the world has a workforce skilled in hardware, software, networking, IoTs, AI, virtual reality, and MR techniques to scale up smart robotics to a global level.

Replacing 4,000 GW of global fossil power opens up a plethora of opportunities. Again, that is where India could play a vital role in this great transformation of the energy sector.

Chapter 11
Rain-Free Agriculture

Agriculture has become extremely challenging in recent times, thanks to chaotic climate change. Rain is not dependable anymore. Flood, drought, and extreme climates are shaking the financial stability of agriculture. According to the World Resources Institute (WRI) the number of people exposed to flood risk is expected to more than double globally by 2030, from 65 million to 132 million people, and to triple by 2050. The World Health Organization estimates 55 million people globally are affected by droughts every year.

This brings us to rain-free agriculture, a trend the world is inescapably moving toward. This method is also known as *vertical farming*, a type of controlled environment agriculture—or weatherproof farming—that consists of fully indoor operations, growing crops on multiple levels solely using electrical lighting, and where you can grow crops without any dependency on seasonal rain. It's exhilarating to hear about such a possibility.

The concept of vertical farming was proposed in 1999 by Dickson Despommier, a professor of Public and Environmental Health at Columbia University. It has quickly earned popularity all over the world as it addresses both water and labor shortages.

Vertical farming is best suited for growing vegetables and grains. This method not only takes us away from rainwater enslavement but also addresses a shortage of workers. The other feature is that the land area needed is ten to twenty times less. For example, a thirty-story building built on a five-acre plot of land can potentially produce an equivalent of 2,400 acres of conventional open farming. This estimate depends on the type of crop. A two-acre vertical farm can produce 720 acres worth of fruits. As vertical farms are staged in covered structures, crops are protected from pests, reducing or eliminating pesticides and consequentially making them organic. This methodology seems to be a perfect solution to all the ills of traditional agriculture.

Once the structure is in place, crops are grown with any of the no-soil farming techniques such as hydroponics, aquaponics, and aeroponics that use materials like polyurethane sponges or biodegradable peat moss instead of regular soil to increase the yield. As the water is fed directly to the roots, 90 percent less water is used compared to traditional open farming. As the fertilizer is fed along with water, its usage is also 90 percent less. As the crops are grown in containers stacked one above the other, the land area use is 90 percent less. This seems like a 90-90-90 savings plan. Since crops are grown in separate containers, different varieties of crops could be grown at the same time. Many factors like humidity, temperature, nutrients, water, and light intensity could be modulated to increase yield and goose up the return on investment.

The time between crops is almost nil as compared to the weeks or months in traditional farming. This is because tractors are needed to recultivate the land, fertilize it, and prepare the soil for the next crop, whereas in vertical farming none of those complexities exist. As most of these vertical farms are being established close to cities, there are ample savings in transportation and warehousing too.

If this is such an awesome technology, then why isn't it happening on a large commercial scale? The snag is the upfront cost. Availability of low-cost structures is key for the success of vertical farming as a considerable investment goes toward the structure itself. Unused, underutilized, or abandoned buildings in cities are the most cost-effective solutions to entrench vertical farming. Even shipping containers, tunnels, and abandoned mines are being used in vertical farming.

But even though there are significant savings in land, labor, fertilizer, and transportation, it's hard to beat traditional farming, however inefficient, because most farmers already own the land, hence no hefty upfront cost. Most wetlands are on the riverbeds, which means there is no water pumping expense. However, the two sharp factors that are tilting circumstances in favor of vertical farming are frequent droughts and labor shortages. In the past couple of decades, climate change has significantly affected the rain pattern. The world is looking for an answer, and vertical farming seems to be one.

In developed countries like the United States, vertical farming is undergoing significant digital transformation to cut labor expenses even further. Artificial intelligence is being employed to monitor and moderate water, fertilizer, and light colors for high yield. Just by manipulating blue, red, and full-spectrum LED lights, they are growing healthy plants. They call it a *light recipe*. By varying the intensity and frequency of light, the color, shape, texture, size, and even flavor of the produce can be changed. They are using AI algorithms to deliver precise nutrients to the plant roots based on the data captured by IoTs.

India is way behind this green, rain-free agricultural revolution. To counter labor shortages, big, rich farmers are still using tractors and tillers for irrigation. A whole new industry has risen to manufacture agricultural equipment to cater to this demand. But what value do agricultural machines add when there is no water? There is a grim shortage of rainwater all over India. Small farmers are grappling to survive droughts, which inflicted financial losses that caused more than sixty thousand farmers to commit suicide every year.

Drought-induced poverty is forcing the mass migration of youth to cities for education and jobs. If this trend continues, it will sooner or later drive all the youths from their villages. In the United States only 20 percent of the population lives in towns, whereas in India more than 65 percent of the population—more than half a billion—still lives in villages. We will surely be witnessing an unprecedented migration in the nation's history, leaving the agriculture industry in a potentially precarious state.

For India to catch up to the rest of the world that's moving toward rain-free agriculture, it unquestionably has to skip the largescale mechanization of open-field agriculture—at least in drought-stricken areas—and jump straight into vertical farming. As more areas face drought in the coming years, they would all go through a similar transmutation of agriculture. This paradigm shift will surely become a template for other drought-stricken nations to derive inspiration from.

Just to give you a feel for this market, the size of the US farming industry is about $130 billion annually; the vertical farming industry—led by Aerofarms, Plenty, Green Spirit Farms, and Bowery Farming, among others—is barely $3 billion. The surface of this mega-industry has not even been scraped yet. Many multimillion-dollar companies are now showing up in this sphere.

In India, agriculture is a $300 billion industry, and vertical farming is still in its infancy. But there is clearly huge potential in this budding industry. Many vertical farming start-ups have already surfaced in India, such as Growing Greens, UGF, Homecrop, Living Food, Urban Kisan, Sure Grow, Kisano, Living Greens, City Greens, Future Farms, and Woolly Farms.

If you look at the global potential of this industry, it's not hard to envision vertical farming businesses springing up all over urban landscapes, fusing urban centers with nature. No need for city dwellers to visit farms to be with nature; farms are coming to them. We're already seeing crops growing in offices, transforming the ambiance of the corporate world. Many companies are already selling small vertical farm units so people can grow fruits and vegetables in their own kitchen. Though these ventures are still small-scale, it's a healthy, inspiring beginning.

Tesla founder Elon Musk has started an agriculture initiative featuring an open-source library that provides plant-specific data, including temperature, moisture, nutrient quantity, type of lighting, and so on. Anyone in the world who wants to get into the vertical farming business can access this data to grow a variety of plants in their own buildings. AI algorithms could also be coded to use this data to monitor water and fertilizer use more efficiently. With this kind of open-source digital recipe, vertical farming businesses could grow crops anywhere in the world, significantly reducing food transportation costs.

In the future, vertical farming would become a hybrid profession that blends the knowledge of both farming and computer software skills. In India, as village agriculture disappears, vertical farming businesses would appear in cities, creating a plethora of high-tech jobs. So as village labor migrates to cities, farming would migrate with them—an enigmatic dual migration made possible only by the power of digitization.

A farmer who migrates from a poor, backward village in India to a booming city could learn computer science and establish a career as a hybrid farmer. What a paradigm shift!

A demographic shift from rural to urban is inevitable as is the evolution of vertical farming. Rather the fret over the loss of village agriculture, embrace the opportunities of high-tech farming. The future is bright.

India not only has a tremendous opportunity in vertical farming, but its IT expertise will also lead the digital transformation of the global agriculture industry. Be it robotics, logistics, transportation, high-tech warehousing, or high-tech retailing, a skilled Indian IT workforce will be a great asset to a digital-hungry world.

Part III
India, the Sleeping Giant

Chapter 12
Waking Up the IT Giant

With more than 400 million youths below the age of fifteen, India is truly a sleeping giant. Even though it has more than 1,200 engineering colleges and more than five million global IT workforce, for India to become the world's leading digital hub, there is a need for a strategic approach designed to fill the skills gap.

Traditionally employers used to hire candidates who were proficient across the board, familiar with multiple fields even if they never used that knowledge in their career. The strategy was to have people who were generalists—Jacks and Jills of all trades. That is changing. Today, employers are looking for specialists who can nimbly provide the skills a particular company needs to succeed in this digital age.

Technology is changing fast. Colleges only provide the fundamentals. Real-world work requires using those fundamentals to create new applications, new solutions, new looking-around-the-corner innovations. The next IT workforce generation must be agile enough to learn new skills and stay current all the time. Their ability to learn skills on the go becomes the determining factor to progress in their career. Indians are very good in this landscape. They have the nimbleness to learn in defiance of poverty, which has brought many successful global careers as CEOs of leading IT organizations.

To produce a vast IT workforce for the world, it is now time for India to focus on expanding the educational opportunities far into the poor villages. The more skilled IT workers India produces, the more it helps other countries implement digitalization. The world is counting on India's youth to fill the IT expertise void that stands in the way of self-reliance.

From my experience here in the United States, the educational standards of India's premier institutions are acknowledged as one of the best in the industry. The Indian workforce has already far exceeded expectations in many innovative, cutting-edge technologies.

However, not all institutions in India are up to global standards. Even though India is currently producing thousands of IT graduates every year, employers are still not finding the right workers they need because they never invested in training the candidates. Instead they are skimming through the traditional education system, which is not focused on digital transformation. That's why there are so many positions companies cannot fill—the current crop of graduates lacks the skill sets needed for the digital landscape. At the same time there is both unemployment and worker shortages in the field of IT.

This skill void can only be filled if candidates are trained the right way. In India, employers in the IT industry need to be vested in education. Instead of scanning through the multitude of candidates who have been churned out inadequately, employers and IT consultancy corporations need to invest in training them to suit the market.

As of today the curriculum in most technical universities is still lagging far behind what the IT industry needs. The market is changing fast, but the conventional curriculum is not catching up. That void can only be filled by the private sector, which knows the market demand and skills needed globally.

Potential employers could train students strategically starting from elementary school, focusing on the latest technologies followed by a couple of years of on-site internship that would bring them up to speed with the global industry standards. India's 400 million youths are spread over six-hundred-thousand villages. If the employers focus on training these exuberant youth, the return on investment would be prodigious when those trained youth work for them.

The stark reason why employers are not investing in school education is because of the long educational wait time, the huge upfront costs, and an uncertain future. But all that is changing now. The future is not uncertain anymore; the whole world is moving toward digitization. With that kind of upheaval in market demand, the upfront cost is justifiable, and the long wait time will become part of the investment strategy.

Viral pandemics are pushing the world toward self-sufficiency via digitization. What the world needs at this stage is an awareness among entrepreneurs about this huge investment potential and India's mammoth human capital. It's time to look at education from an investment perspective rather than just philanthropy.

The day when entrepreneurs invest in building smart schools to produce a new breed of smart graduates braced with smart technologies, India, the sleeping giant, will really awaken.

Is the world ready to tap into India's potential?

Chapter 13
Waking Up the Business Giant

India is poised to take an entrepreneurial leap to become self-reliant. But many age-old companies in India are nowhere near digitization. Moreover, many of those companies are resolutely reluctant to change as they have already invested heavily in their existing old technologies. Decommissioning that old clutter and revamping it with the latest technologies is prohibitively expensive. That's where innovative tech start-ups could give them a run for their money—the reason why innovative tech start-ups are notoriously nicknamed *disruptive innovators*. They disrupt the status quo.

According to the Indian ministry of commerce and industry: "India has the third-largest start-up ecosystem in the world; India had about fifty thousand start-ups in 2018; around 8,900–9,300 of these are technology-led start-ups, 1300 new tech start-ups were born in 2019 alone, implying there are two to three tech start-ups born every day."

Although the Indian government is encouraging start-ups through various incentives, many start-ups fail during their first few years. About 20 percent fail in their first year itself, 30 percent in their second year, and another 40 percent during their fifth year. Entrepreneurship is not a bed of roses. There is intense competition, and the market is brutal. The failure rate of start-ups globally is close to 90 percent. Let's explore why so many start-ups fail and how the successful ones navigate that brutal competitive landscape under the umbrella of digital transformation.

Many start-ups fail even if they have brilliant business ideas, astute market analysis, and abundant funds. All the failures have one thing in common—a lack of perpetual innovation. Companies need stout leadership to constantly innovate the business. If you look at any billionaire investors in the world today, they all invest in the leadership, not in the business. That's one of the reasons why CEOs are paid so handsomely throughout the world. They're not hired to run the business but to steer it constantly to respond to the market.

Amazon for example started as an online bookseller. Then it constantly innovated and stayed ahead of the competition. Today it's not only the largest online seller in the world but also a global leader in cloud technology called AWS.

If you look at the world of start-ups, there is tough competition. All start-ups work hard, they do proficient market analysis, hire a highly skilled workforce, raise copious funds, and time their product launches very well.

However, as time progresses some start-ups catch on while others fall off. All successful ones have one thing in common: reinvention. As Tesla CEO Elon Musk says, if a company can't constantly innovate, it is bound to fail. The same is true with key employees of companies; if they cannot show a passion for innovation, their employment will be at stake.

Continuous innovation is not easy. That's why many start-ups in today's world try to quickly sell their business after establishing a little market presence. In the same breath there are also larger companies that are constantly buying smaller businesses to expand their market share. If a new successful start-up shows up, larger ones will bid to buy it. As you can see, there is both entrepreneurial zeal to innovate and an equally strong gusto to chop off the competition. In this fishpond, bigger companies become bigger and bigger, gulping up smaller ones. But it does not mean big companies are safe either. A full monopoly over the market does not guarantee them success forever. There is always competition lurking in the corner, and they have to keep pressing that peddle of innovation.

This is a daunting challenge for budding entrepreneurs. As a start-up they begin with a zest to innovate, and then they become a meal for hungry wolves. This landscape is brutal for start-ups to survive.

But digital transformation is giving this cruel market a new look. As we all know, AI is taking the world by storm. Traditionally our engineers used to spend months or years designing a product. Now, AI can do the same in seconds. A year's worth of hard work is now reduced to seconds. Not only that, AI can give them not just one design but hundreds of designs to choose from. With that kind of power in hand, even multinational companies are threatened by garage start-ups.

AI is being used in all facets of business, be it manufacturing, supply chain, marketing—you name it, there it is. It's not easy to tell which company is really doing a professional job, as AI is doing its charm behind the scenes. These cutting-edge technologies have changed the competitive landscape completely. Small start-ups can now easily compete with the larger ones now. For example, Tesla showed the world the importance of innovation. In just a few years, it changed the automotive industry. Tesla's electric cars are so successful, car companies that had never thought of producing electric cars are now scrambling to get back their market share and are investing heavily in electric automotive technology. Just one start-up has changed the entire industry's outlook.

In India, traditional mega-companies will fall quickly if they are not on par with digitization. Most developed regions like the United States and Europe went through massive globalization, with AI, robotics, and machine automation the end results. On the other hand India missed all that globalization for decades. Many Indian industries are lagging behind considerably in the latest technologies. That is where the opportunity lies. The future Indian digital transformation market is immensely large.

It's estimated that the Indian domestic digital transformation market is expected to reach $700 billion by 2025. Technologies like AI, IoT, big data, cloud, blockchain, augmented reality, and virtual reality are expected to lead the transformation market in the areas of smart manufacturing, smart logistics, smart transportation, smart energy, smart cities, smart healthcare, and smart agriculture.

India's digital battleground is quite unique. Instead of mega-companies threatening start-ups, innovative start-ups are threatening mega-companies. However chaotic that may look like, in this digital battle, the only strategy that survives is perpetual digital transformation. The start-ups are waking up the sleeping giant corporations. From that perspective the more start-ups there are, the better it is for India. This brutal digital war unquestionably offers a huge digital market and creates a plethora of opportunities for the Indian IT workforce.

While India is attempting *Atmanirbhar Bharat—self-reliant India*—and reeling through its own stiff digital battle, the world is counting on India to offer its IT services to help them become self-reliant as well. India indeed has a dual role. For India this is an opportunity, a responsibility, a technological challenge, and an ethical obligation all bundled together.

This is a rare moment for India to be vested in education. A rare moment indeed.

Part IV
Tech Primer

(A dig at technologies)

(Optional Read)

Chapter 14
IoTs

This last part of the book explores some of the leading-edge technologies that are behind the rise of the smart world. The first is IoTs.

IoTs are phenomenal entities. In today's world any digital device that connects to the Internet is termed a smart device. In essence all smart devices—smartwatches, smart garage doors, smart TVs, smart refrigerators—are IoTs. Any gadget can be made smart by giving it the capability to connect to the Internet. A device cannot be an IoT, no matter how sophisticated it is, if it is not able to connect to the Internet. The reason for emphasizing an Internet connection is that it gives the device a unique identity in the digital world. We call this unique identity an Internet protocol or IP address. An IoT can't connect to the Internet unless it has an IP address. There could be IoTs on a local cloud system or fog and may or may not connect to the Internet all the time. However, in the broader sense IoT gets its significance if it has an Internet presence.

In smart homes AI needs a lot of data to learn and become a useful tool. For instance, if you want AI to prepare coffee when you wake up, switch off the lights when you leave the house, or start the furnace and keep the house warm before you arrive, AI needs data to learn your lifestyle. All smart gadgets in the house keep tabs on you. That is how AI learns your habits and lifestyle to serve you better. Whenever you shop online, watch a movie, chat on Facebook, Google a word, or ask Alexa a question, the AI learns. There is not just one AI; pretty much all smart gadgets are run by their own AI. We have come a long way and are knee-deep in the smart world. There is no going back. Some may look at it as an invasion of privacy, but those thoughts are a little late in the game.

Here is a good example of IoT in healthcare. Say a patient is remotely monitored with a wearable device (IoT), and their heart rate increases to an unsafe level. The real-time IoT data has to be analyzed, the abnormalities need to be identified, and the hospital staff has to be alerted. If this responsibility is assigned to an individual who is monitoring patients remotely, reviewing data from thousands of patients in real-time is impossible. Even setting up software with rules is time-consuming and fraught with errors. That's where AI helps us out. A well-trained AI algorithm can analyze the data in real-time and flag anything outside the norm.

The company Somatix has come up with an AI-powered wearable SafeBeing, a passive and remote monitoring platform used to monitor daily activities such as mobility, fall detection, hydration, activity patterns, wandering, and more to provide insights and predictive analytics and enabling caregivers to detect changes in a patient's condition thereby reducing hospitalizations and improving quality of life.

IoTs around the world are producing a treasure trove of machine data on traffic, weather, crime, healthcare, supermarkets, industrial robotics, smart homes, and so on. AI is taking advantage of big data and is evolving to help us with sales projections, crime prediction, weather forecasting, object, face, language, and voice recognition, and even playing chess.

The RFID Riddle

It's worth discussing radio frequency identification (RFID), as it coexists with IoTs in many places. RFID is a digital barcode (microchip). In a regular bar code, the data is static; it can't be updated. In an RFID system, data can be updated. RFID tags are good for storing vital information like expiration date, manufacturing date, package destination, and so on.

The RFID tags or microchips can be attached to manufacturing products in factories to track their progress through the assembly line. The RFID-tagged pharmaceutical products are very popular. RFID tags can even be used with livestock and pets for their identification and tracking.

As RFID tags do not have the capability to connect to the Internet, they need a reader to access and send their data to the Internet, making it an IoT. To understand the usefulness of RFID, let's take the example of the smart refrigerator. Say you bought a carton of milk with an RFID tag and stored it in your smart refrigerator. The RFID reader in the refrigerator reads the tag and sends its expiration date to your phone app via the Internet. However, RFID can't sense the weight of the milk carton and hence can't send you information about how much milk is still left in the carton. RFID can only give static data and not dynamic data as it is not a sensor.

RFID existed well before IoTs came into the market. They're gaining more importance in today's logistic world as they produce vital data.

RFIDs can cost anywhere from a few cents to several dollars, depending on the size and range of data transmittal. They can be as small as postage stamps or as big as your car transponder. RFID application is widespread in credit cards, hotel room cards, car keys, car transponders, passports, ID cards, driver license cards, bicycles, pets, and even hospital patients. They are pretty much everywhere. RFID is useful in pharmacies to manage inventory.

An RFID reader can simultaneously read several hundred tags. It can hang from the ceiling and read all the tags in the warehouse and send the inventory list to the Internet.

Pretty much all trucks have RFID. As they go from one warehouse to another, they are read and tracked. Many cars have RFID transponders that can be used in tollways and parking lots to open the gates.

Although RFID is useful, the technology has limitations. Data security is a major issue as RFID can reveal our secrets. Say you bought an expensive Rolex watch; the RFID tag on the watch could be broadcasting as you walk down the street. A smart thief can figure it out. These kinds of security issues are being ironed out.

There are also some technical problems with RFID. Since an RFID uses radio frequencies, it can jam when using overlapping frequencies. That could be life-threatening in a hospital. RFID readers are prone to collision issues when signals overlap. However, anti-collision protocols are being developed to counter that. The bottom line is that technical challenges are not showstoppers; they only make the system better. Amid all the technical limitations, RFID has become an extended arm for IoTs to capture data.

The takeaway from this discussion is that IoTs and RFID are like eyes and ears for AI to make decisions. Whenever you hear the words like *big data*, remember IoTs and RFID; they are the ones producing it along with human-generated social media data.

A Word about Big Data

Every minute about three million messages are posted on Facebook, fifty thousand pictures are shared, half a million tweets tweeted out, and forty million texts are sent. Every time we shop, we spawn tons of data tied to banks, retailers, and industries. This mountain of data is given a name: big data. The blast of big data is so huge it's humanly impossible to make sense of it. This level of complexity and volume of data gave birth to software like data analytics tools that extract, analyze, and interpret data in a way that is useful and applicable.

However, even the most sophisticated data analytics software has limited capability and will only do exactly what it's coded for. Big data has innumerable data patterns hidden in it, and it's not possible to code software to unearth all the individual combinations. So AI emerged as a natural solution to pull meaningful information out of big data. AI can identify observable patterns and predict things without having explicit pre-programmed rules and models.

For example, if you want to know what city has more crime, a simple database query can figure it out, or a simple data analytics tool can answer that instantly. However, if you want to know what specific locality is going to get hit by a criminal gang in the next twenty-four hours, you have to go through numerous data sources such as Facebook postings, tweets, texts, Google searches, local demographics, and criminal reports, accumulating innumerable data points. It would be humanly impossible to connect all this structured and unstructured data to predict a pattern of criminal acts. It's like looking for a needle in a haystack of needles. You need an AI that correlates this massive data in a useful way.

To find those hidden patterns, AI is trained to predict some past events that have already occurred based on past data. If it makes a mistake, it reevaluates and predicts again, learning iteratively over and over again until it becomes a useful tool. How well AI predicts depends on the quality of data. Garbage in, garbage out is appropriate here. The quality of big data is as important as skillfully coded AI itself. If you want AI to predict anything, the data needs to be vast, varied, high-quality, and as real-time as feasible.

Broadly speaking, there are two kinds of data sources: human-generated and machine-generated. Human-generated data includes social media, blogs, Google search, texting, tweeting, online shopping, and so on. Machine-generated data comes from IoTs and RFIDs.

Human-generated data is increasing every day as more humans are joining the Internet from the developing and underdeveloped world. Once all seven billion people on the planet gain access to the Internet, the growth of human-generated data will flatten out. However, machine-generated data will continue to grow unabated as machines around us keep increasing, moving us toward a machine world, making AI an inevitable part of our life to make sense of this big data.

Chapter 15
Artificial Intelligence

Artificial intelligence has really fascinated the world. While humans are still early in understanding human intelligence, they have the audacity to build artificial intelligence.

Being in the IT industry myself for more than two decades, my fascination with this new software kept growing as it became more widespread every day. It's just unbelievable that this weird term that made its academic appearance recently already has a career path in its name. I began to wonder if all the hype was because it sounds so cutting edge or if it was because AI has true value. I started my own reconnoiter out of unquenchable curiosity. The more I dug, the more captivating I found it.

When it comes to understanding computer programs, there is nothing equivalent to wetting our hands. So I started to learn AI by coding a self-driving virtual car on my computer that finds its way to a destination with the shortest route. After all the hard work and finally completing the python code, a small, ant-like bug began to crawl on my computer screen, going all over in random directions. That was my AI algorithm.

Amazingly, when I keyed in the start and end positions, this little creature that was crawling all around the screen slowly changed its course and began to move around the starting position for a while, and then it crawled toward the end position. Once it reached the end position, after maybe ten or fifteen back and forth movements, it learned the quickest way to move between those two points. However, even after finding the shortest route, it did not give up its habit of sniffing around to find potential shorter routes, if any.

That constant sniffing gives AI the special ability to learn from its own mistakes just like humans do. It was just an incredible experience. It's not a typical software that obediently follows the hard rules; this was an ever-learning software. What was moving like an ant all around the screen now began to look more like a commuter car between downtown and the airport. I had built software that learns on its own In my career as an IT professional, I am used to giving commands to software to do tasks for a specific result. I had never written software that could act on its own to achieve a specific result. It changed my perception of AI forever. So I started learning more about this amazing human creation.

You may be wondering how this software learned the shortest route. The simplest explanation is that it's coded in such a way that the behavior of the bug is dependent on what is called a reward point. Every time the bug moves closer to the destination, it gets a reward point. Every time the bug moves away from the destination, it loses a point. The computer program has been coded to bag as many rewards as possible.

How does it know which way it has to move to get closer without GPS or any direction sensor? It uses a heuristic technique—essentially trial and error—to compute the distance. That data is then automatically stored in a memory matrix called a learning library. That's why it initially goes in random directions to compute the most optimal way. This computation of optimal path is called training the AI, a process that is highly central processing unit (CPU) intensive. Once trained, the data is stored in the learning library for later use, and it keeps updating it as well. This unique learning capability is the reason it's becoming so powerful and the darling of the world.

A heuristic technique is not the only one used in AI. There are tons of them, and they are not new. These are age-old ideas and mathematical models that were never tested on computers because they were dead-slow in previous decades. Now with faster computers, it's no surprise old AI concepts are being rediscovered.

Another popular AI technique that is widely used is called artificial neural networks or simply neural net (NN). This is one of the top-notch techniques modeled after the neural structure of the human brain. The disparity is that NN might use a few hundred neurons, whereas the neural structure of the human brain has approximately one hundred billion neurons. It's obvious that humans have a better capability with things like face, object, voice, and natural language recognition, as well as many more complex qualities that machines don't have yet. However, NN has the capability to outpace human neural networks very quickly as it takes only some more CPUs—popularly called chips—and few more gigabits of memory to add more neurons. The truth is the capability of NN is boundless.

Now let's look at the structure of a human neural network a bit more closely from the perspective of data flow. In human brains every neuron is connected to a bunch of other neurons. The output of one neuron becomes the input to other neurons, meaning the signal coming from one neuron is captured by other neurons connected to it as their source of information. Based on multiple inputs, a neuron generates a new output that can be used again as input to other neurons. This goes on like a crazy maze in these one hundred billion neurons.

Though it looks complex, they all follow a distinct rule. The strength of input signals decides the output from a neuron. If a neuron gets high-strength signals from all its inputs, its output will be equally strong, which will be passed on to the next neuron. For example, if by accident touch a hot plate, all the nerve endings on the hand will send out strong signals to the brain, which cause the brain to act, and your hand is pulled back as a reflex action almost before you are aware of it. The brain processes the inputs and acts on them based purely on the strength or weakness of the signals. The same technique has been adapted in NN. Furthermore, each artificial neuron is designed to use a mathematical function to determine its own output.

Here is an example of the object recognition capability of NN. Let's say you have a basket of random objects, and you want AI to recognize them. In reality NN doesn't know a thing about any object in the basket. When you show AI an object like a toy bird, the camera image of the bird is broken by the NN into a number of individual images. Each image is fed to a different neuron that gives its own weight to the image based on its own calculations.

Some of these images could be very prominent features for identification, while others may not be. For instance, the beak is a prominent feature to identify a bird. As the NN is not initially aware of these prominent features, it gives random weight for their prominence. Discernibly, the prediction would be random too. If the prediction is incorrect, then weights are readjusted. This happens over and over again until the NN learns all the features that are useful for identification. It will eventually determine which features of the bird are more prominent. The more data you use to train NN, the closer they get to the right answer.

After training, when a photo of a new bird is shown, the NN will give its best answer. The accuracy of the answer depends on the volume and variations of the data that was fed during training. In other words the quality of data matters. If there is a weird bird that is confusing to the human eye, AI might also falter if that kind of data is not in the database and was not fed in during the training.

Obviously, if you show the image of an elephant to the bird-trained AI, it won't know what it is because it would need specific training on elephants. You can't possibly train AI on every species, so this is where big data comes in. If the information about all animals is available in a database, fully identified and tagged, then the AI can skim through the database and learn from it instantly.

Now let's explore some interesting AI applications we have in the market today.

AI in Movies

The entertainment industry is attracting AI to streamline the movie-making process to give audiences more bang for their buck.

In 2016 an AI software, IBM Watson, created a ten-minute trailer for the movie *Morgan*. Watson went through the movie and selected a few emotional movements like love, hate, anger, and horror, then made a few short clippings for the director to choose from. (To see the trailer Google *AI Morgan trailer*; it's enthralling to watch.) AI not only saved millions of dollars, but it also introduced its own machine ideas, surprising even the creators of AI.

Similar attempts are going on in Japan. McCann Erickson of Japan introduced a new AI "creative director" named AI-CD ß that made a commercial for Clorets mints. The AI was given a decade's worth of commercial ads to learn from. (To see it in action, Google *AI-CD beta*.)

Though AI has not been fully established in the film industry, it's at the doorstep.

AI in Sports

AI has entered Wimbledon. IBM Watson uses crowd noise, social traction, facial recognition, and sentiment analysis of players to generate short video highlights. So a video editor will no longer be needed to cut and edit to put a highlights package together.

AI in Chatbots

Chances are you already have some sort of chatty virtual assistant or chatbot at home. Alexa, SIRI, Google Home, Cortana, and so on are all AI-driven chatbots. Although the AI in these devices is still in an infancy stage, they are pretty amazing. They understand many languages and multiple ways of asking questions, so you don't need to worry about remembering exact phrasing for commands. Google Home is loaded with predictive features. Microsoft's Cortana is great for location-based reminders.

Most shopping websites are now chat enabled. Companies train these chatbots—literally chat robots—before putting them live for customers to use. During training they use the most frequently used questions and relevant answers, creating an answer bank. However, if the question that a customer asks is outside of the answer bank, then it will either give an irrelevant answer or a totally wrong answer. However, AI would learn every time you chat with them.

These chatbots listen to you and observe your behavior. They learn about you every time you tweet, watch a movie, listen to music, do a Google search, or make online purchases. As they learn they serve customers more efficiently. This may look more like an invasion of privacy, but that is what the world chose.

Here is a simple litmus test to see if the chatbot that claims to be AI really is or not. Ask the virtual assistant some sample questions and note the responses. If you repeatedly ask these questions over a certain period of time, maybe a few weeks or months, and the answer is the same, most probably it is not yet using AI. On the other hand, if this gadget is learning about you and is giving you a more useful response each time, then in all probability it could be using some sort of AI.

Two-Day Product Delivery

Major retailers now do purchase prediction. Most of the time when you browse for a product, the AI in their websites identifies you as a potential customer and emails you special offers. They can send you coupons, offer you discounts, and target you with advertisements while also stocking their warehouses that are close to you with products that you're likely to buy. That is how your product is delivered the next day.

The two-day product delivery strategy is becoming smarter and looks more like fire station services. For fire engines to get to the fire within a specific time frame, stations are strategically located throughout the city. A similar system has been strategized for two-day product delivery, which is why Amazon has been building warehouses all over the United States. Each warehouse services a specific area, and the AI knows the needs of that population, so each warehouse is uniquely stocked to cater to a specific demographic population. For instance, if the locality has more elderly, then more caregiving products are stored. If the locality has more kids, more toys are stored. Even the dress sizes they store are based on the demographics. The goal is to minimize the inventory and deliver quickly. When customers browse for products online, AI knows which products will be in the greatest demand during the coming holiday season and stored accordingly. All these decisions need a lot of data, and analytics in real-time is done every minute of the day. Behind the scenes AI would be making these decisions and constantly learning from its mistakes.

AI in Humanoids

The robot Sophia is a social humanoid created by Hong Kong-based Hanson Robotics. Sophia was activated on April 19, 2015. This robot imitates human gestures and facial expressions. It can answer questions on certain predefined topics. This humanoid is designed to be a conversational companion for people in retirement communities, hospitals, and nursing homes. It is also capable of entertaining crowds through social conversation. This humanoid is on YouTube, and it's captivating to watch the facial expressions on this robot.

The AI algorithm was designed by SingularityNET, which aims to foster an open market for AIs.

AI in Policing

The American company ShotSpotter has developed a technology to help identify gunshots. It installs sensors in tall buildings in cities, and when a gun is fired, these sensors capture the sound and vibrations and mutually exchange data to pinpoint the exact location of the gunshot and alert the police. This is not only saving lives but also reducing the time and effort to respond, eventually saving even more lives.

Another interesting innovation comes from an American company called PredPol, which can detect crime before it happens. They look at history and figure out the crime trend and identify potential areas and possible time of day. Though they can't predict exact location or exact time, their predictions would help police be on alert in those targeted during a specified window of time, saving lives, time, and money.

The Chinese company Hikvision uses video surveillance cameras to read license plates, use facial recognition, and even detect unattended bags in crowded areas and alert law enforcement officers.

Another Chinese company, Cloud Walk, is going one step ahead. It is using facial recognition software to identify suspicious and unusual body language that could trigger a security alert. AI is evolving faster to counter crime.

AI in Recruiting

If you are looking for a job, chances are you will be interacting with a virtual assistant, most likely driven by AI. You better watch out. These AIs learn a lot about you and stores all that data digitally, so your profile will be available to all potential employers.

Employers should also be equally concerned as their profile will also be available to all potential candidates. If the employer is racially biased, it will show up in the recruitment history.

You can't avoid this route. So be prepared to face this digital challenge and be attentive to what you give out. Since you will be interfacing with a machine instead of a human, it could be a new experience. You may ask the wrong questions or give the wrong answers to AI. Everything is digitally recorded.

One fascinating thing that has happened to the recruitment process is that biases are slowly being extinguished. The AI software scans through the résumés of potential candidates without considering gender, age, and name to eliminate biases that could steer the focus away from promising candidates. It also flags missing information from the résumés that the employer is looking for.

When candidates apply for a job, even before the résumés reaches the employer, AI-driven software will communicate with candidates not only to collect more information and do prescreening tests but also to share employer information. This helps the candidates as well by knowing if they are looking at the right employer before attending any further face-to-face interviews to avoid wasting time for both. Employers love this approach. AI asks challenging questions and looks for effective answers.

Many recruiting AIs are becoming proactive and going one step further in compiling a list of potential candidates not currently in the job market but who may seek employment in the future. AI can identify which candidate would be likely to change jobs because of a company merger or layoffs. And if a candidate updates their LinkedIn profile, it's a sign that the candidate could be looking for a new job.

Hiring is now popularly referred to as talent acquisition. There are plenty of talent acquisition companies like LinkedIn, Indeed, Glassdoor, Dice, Monster, Craigslist, Plaxo, Jobster, and so on. There are also many companies that are developing talent acquisition software like Ideal, Avrio, Entelo, Engage Talent, Paradox Olivia, and Mya Systems. You may visit their sites to learn how they are developing and using AI to acquire talent.

AI in the Stock Market

Picking promising stocks every day is arduous for portfolio managers and needs a lot of experience and skills. They have to scan data coming from news media, social media, blogs, company announcements, consumer confidence reports, real-time volatility in the stock market, etc. The amount of data is so exhaustive it is becoming hard to analyze and arrive at fortune stocks to invest in. This is where machine learning is picking up steam.

The American company Kavout is engaged in building deep learning algorithms to assist individuals in trading. People can now hire AI instead of humans for trading.

AI is also influencing algorithmic trading and high-frequency trading systems. Many robot advisors are in the market today. Most hedge funds and financial institutions have their own version of AI.

As the data is so vast and varied, AIs of different companies analyze them in different ways. There is intense competition to build the best AI for the stock market. The race is on; the winners are the individuals like us. Skilled jobs are plenty in this ever-evolving industry.

AI still needs babysitting in many ways, as any software glitch could wipe out portfolios. The computer glitch of 2012 has left a bad taste in everyone's mouth. Although AI is threatening traditional portfolio management jobs, the financial market still needs people who can understand both AI and business at the same time. Portfolio managers have to learn to live with AI. AI is both a job threat and a job opportunity at the same time.

AI in Rescue Operations

Do you know there are unique robots that look like and act like animals? For example, eMotion Butterflies from the company Festo fly like real ones. Their dragonfly-like BionicOpter robots fly and glide in the air. Their bionic ants are the size of a hand. Their AquaPenguin can swim in water as well as look and act like real penguins. Their bionic kangaroo looks like and jumps like a real kangaroo. There are captivating videos online that are worth watching.

The Defense Advanced Research Projects Agency (DARPA), a wing of the US Department of Defense, has created a variety of incredible robots. It's riveting to watch these robots in action on YouTube videos.

These amazing machines are a blessing in disguise for disaster response. They can fly, crawl, or walk into places humans can't. They can operate in environments where humans can't even survive. During disasters like hurricanes, tornadoes, floods, forest fires, nuclear leakages, and earthquakes, they bring data of immense value. While some of them can collect data, others can help in real rescue operations.

AI in Drones

In a captivating TED Talk worth watching, Professor Vijay Kumar from the University of Pennsylvania demonstrated tiny drones performing a series of intricate maneuvers, flying through confined spaces without colliding or interfering with each other. These smart robots/drones are capable of aiding in construction, shipping, and even responding to emergencies.

AI in Weather Forecasting

Have you heard of hindcasting? We know forecasting, but what is hindcasting? Forecasting past events using past data is hindcasting. This is how AI is trained. AI is trained to hindcast first before forecasting. All AI-driven climate models that are out there today are trained with historical data dating back many decades. Along with extensive real-time data, the more historical data we provide better would be the forecast.

The Weather Company, an IBM business, is aggressively improving its AI-driven weather forecasting model called Deep Thunder, which handles almost four hundred terabytes of data every day, generating tens of millions of forecasts around the globe at fifteen-minute intervals.

Panasonic is yet another company that is deep into improving weather technology. It has started a similar global forecasting system called Panasonic Global 4D which predicted Hurricane Irma more accurately than other models.

Threats to AI

There are two broad categories of AI. The first is a hard-coded slave category that mimics human actions. This category has many names: basic AI, reactive AI, weak AI, or narrow AI (ANI).

The second category is widely called general-purpose AI or AGI. This can learn on its own and improve over time, a kind of self-learning. Most AIs that we have today are hybrids with both rigid ANI features and some AGI learning capability.

We shouldn't get too bogged down with these academic terminologies of AI classifications; however, we need to know their distinction to understand the technological threats they are facing.

Some AI experts think that it could be dangerous that we still have many rigid ANIs that are not learning from their mistakes. Anyone can hack ANI to knock out our electric grid, damage nuclear power plants, and misdirect robots causing global-scale economic damages.

There is an urgent need to develop smart AIs that we all can depend on. Till then we have to be on constant vigil to protect ANIs from outside attacks. We shouldn't get too excited if some utility company adopts an AI. If it's more of a basic ANI with little or no self-learning capability, it could be doing more harm than good as it can't figure out the attackers. It is better that we understand this AI categorization to know the threats we are facing.

Most AI software is proprietary, which means we don't know how they are coded. The AI coded by Google may be entirely different from AI coded by Facebook or Amazon. Aptly, open-source is beginning to take shape to exchange AI codes and its faster evolution.

Until ANI evolves to a safe zone, there will be a great demand for anti-viruses to thwart ANI hackers. These are the areas of new opportunities. There is also an effort to build a library of ANI talents to evolve ANI quickly. As this library and the open AI platform widens globally, future opportunities in the AI world would be unprecedented.

Careers in AI

Many AI trend trackers are estimating that more than half the current United States workforce will be at risk in the next decade or so. They are also predicting a plethora of new jobs in AI-related fields.

AI uses many machine learning techniques such as heuristic, deep Q-learning, reinforcement learning, neural networks, and many more coming. Most AI systems are written in Python and R languages, although AI can be coded in any language. In the future as more powerful languages are introduced, programmers would make that shift. There are many AI courses available online for serious learners and for academic enthusiasts. Those who are starting new careers and willing to get trained in cutting-edge technologies shouldn't ignore AI.

I hope this chapter gave you a quick glimpse into the world of AI. The sky is the limit when it comes to learning cutting-edge technologies. Keep on learning and help the world improve the technologies. That is the best gift you can offer to the next generation.

Epilogue
Why a Smart World is a Better Place to Live

Although the smart world is a remarkable human creation, it will not come without collateral damages, such as more unemployment among the elderly as many won't be able to keep up with the tech world.

There will be an equal number of high-tech jobs for youth who could join the workforce as teens or young adults. But they could fall out of the workforce early.

We are already beginning to see both trends.

With today's increasing longevity, it's a huge burden for governments to support the jobless for decades. The good news is that the cost of food, housing, healthcare, transportation, and many other services will be much less in the smart world. All the governments around the world would be able to look after their citizens fairly well. When basic needs like food, shelter, and medication are guaranteed, there will be more tranquility and peace in the populace.

Everyone has a stake in the eventual smart world, and both rich and poor will win in a smart world. Whether companies make or lose money, whether individuals succeed in their career or fail, at the end of the day, there will be food on the table. Everyone will have a roof over their head, and above all the elderly are taken care of. That's what makes the world truly smart. Would a smart world be a better place to live? You bet.

Technologies have never regressed; they have always evolved. However, their growth may slow down if there is not enough skill in the world. This is a rare opportunity and the perfect moment for India to be vested in education.

Acknowledgments

Grateful acknowledgment to the many international organizations that provided useful quality data for public research, including the World Health Organization, National Center for Public Policy and Higher Education, www.House.gov, www.cbo.gov, www.eia.doe.gov, www.nasa.gov, www.ers.usda.gov, stanford.edu, forbes.com, ibm.com, dhs.gov, industry40summit.com, darpa.mil, epa.gov, nature.org, and www.cms.gov.

Many thanks to my family and friends for their helpful comments, advice, and support.

Ravi Amblee is the author of THE UGLY FIGHT, the award-winning book that deliberates on artificial intelligence as a weapon against climate change. After completing his studies in artificial intelligence at MIT, he is deeply committed to climate advocacy promoting action against global warming employing artificial intelligence. He has a master's degree in mechanical engineering and an IT career spanning over twenty years. His articles have appeared in various newspapers, including the DECCAN HERALD and THE HINDU, and OUTLOOK INDIA magazine. He has answered more than three hundred questions on Quora about ways to make technology work for humanity to make the world a better place for all of us and has a readership of fifty thousand-plus. Ravi lives in Chicago, Illinois, with his family.

Index

Bibliography

"Artificial Intelligence Replaces Physicists." The Àustralian National University. Last modified May 16, 2016. http://www.anu.edu.au/news/all-news/artificial-intelligence-replaces-physicists

"Artificial Intelligence Takes on the Stock Market." BBC (video). 01:21. February 10, 2016. http://www.bbc.com/news/av/technology-35405336/artificial-intelligence-takes-on-the-stock-market

"Artificial Intelligence." Wikipedia. Accessed July 12, 2018. https://en.wikipedia.org/wiki/Artificial_intelligenc e

Baker, David R. "Robots Cut Solar Construction Costs." San Francisco Chronicle. June 17, 2013. https://www.sfchronicle.com/business/article/Robots-cut-solar-construction-costs-4604343.php

"Beijing Uses Machine Learning and Big Data to Target Pollution Controls." Apolitical. Accessed July 12, 2018. https://apolitical.co/solution_article/beijing-uses-machine-learning-big-data-target-pollution-controls/

Bohannon, John. "A New Breed of Scientist, with Brains of Silicon." Science. July 5, 2017. http://www.sciencemag.org/news/2017/07/new-breed-scientist-brains-silicon

Callaghan, Greg. "Can Swarms of Seed-Bearing Drones Help Regrow the Planet's Forests?" Sydney Morning Herald. August 26, 2017. https://www.smh.com.au/lifestyle/can-swarms-of-seedbearing-drones-help-regrow-the-planets-forests-20170823-gy2ei5.html

Clark, Jen. "What Is the Internet of Things?" Internet of Things Blog. November 17, 2016. https://www.ibm.com/blogs/Internet-of-things/what-is-the-iot/

Conway, Erik. "What's in a Name? Global Warming vs. Climate Change Global." NASA. Accessed July 12, 2018. https://www.nasa.gov/topics/earth/features/climate_by_any_other_name.html

"DARPA Robotics Challenge (DRC) (Archived)." Defense Advanced Research Projects Agency. Accessed July 13, 2018. https://www.darpa.mil/program/darpa-robotics-challenge

"Deep Convective Clouds and Chemistry Experiment (DC3)." The National Center for Atmospheric Research. Accessed July 12, 2018. https://www2.acom.ucar.edu/dc3

"Deep Thunder Now Hyper-Local on a Global Scale." IBM (blog). June 15, 2016. https://www.ibm.com/blogs/research/2016/06/deep-thunder-now-hyper-local-global/

"eMotion Butterflies." Festo. Accessed July 13, 2018. https://www.festo.com/group/en/cms/10216.htm

https://www.nature.org/greenliving/carboncalculator/index.htm

https://www.research.ibm.com/green-horizons/interactive

https://www.transcriptic.com/ "Internet of Things." Wikipedia. Accessed July 12, 2018. https://en.wikipedia.org/wiki/Internet_of_things

https://www.c2es.org/content/wildfires-and-climate-change/

https://nvlpubs.nist.gov/nistpubs/SpecialPublications/NIST.SP.1215.pdf

https://spectrum.ieee.org/automaton/robotics/industrial-robots/robotinstalled-solar-panels-cut-costs-by-50

https://www.start-upindia.gov.in/content/sih/en/international/go-to-market-guide/indian-start-up-ecosystem.html

https://censusindia.gov.in

https://somatix.com/solutions/#contactus

"IPCC Fourth Assessment Report: Climate Change 2007." Intergovernmental Panel on Climate Change. Accessed July 13, 2018. https://www.ipcc.ch/publications_and_data/ar4/wg1/en/ch7s7-3-2-2.html

Kumar, Vijay. "The Future of Flying Objects." Ted (video). 13:10. April 2015.

https://www.ted.com/talks/vijay_kumar_the_futur
e_of_flying_robots

Mahtanim, Shibani and Zusha Elinson. "Artificial
Intelligence Could Soon Enhance Real-Time Police
Surveillance." Wall Street Journal. Last modified
April 3, 2018.
https://www.wsj.com/articles/artificial-
intelligence-could-soon-enhance-real-time-police-
surveillance-1522761813

Marr, Bernard. "What Is the Difference Between Artificial
Intelligence and Machine Learning?" Forbes.
December 6, 2016. https://www.forbes.com

McCarthy, John. "What Is AI?/Basic Questions." Accessed
July 12, 2018. http://jmc.stanford.edu/artificial-
intelligence/what-is-ai/index.html

"NOAA to Develop New Global Weather Model."
National Oceanic and Atmospheric Administration.
July 27, 2016. http://www.noaa.gov/media-
release/noaa-to-develop-new-global-weather-model

O'Reilly, Lara. "A Japanese Ad Agency Invented an AI
Creative Director—and Ad Execs Preferred Its Ad
to a Human's." Business Insider. March 12, 2917.
http://www.businessinsider.com/mccann-japans-
ai-creative-director-creates-better-ads-than-a-
human-2017-3

"Quest Carbon Capture and Storage." Shell Canada.
Accessed July 12, 2018.
https://www.shell.ca/en_ca/about-us/projects-

and-sites/quest-carbon-capture-and-storage-
project.html

"Radio Frequency Identification (RFID): What Is It?"
Department of Homeland Security. Last modified
July 6, 2009. https://www.dhs.gov/radio-
frequency-identification-rfid-what-it

"Radio-Frequency Identification." Wikipedia. Accessed July
12, 2018. https://en.wikipedia.org/wiki/Radio-
frequency_identification

"Safely Storing Carbon Dioxide." Chevron. Accessed July
12, 2018. https://www.chevron.com/stories/safely-
storing-co2

Shaw, Darren. "How Wimbledon Is Using IBM Watson AI
to Power Highlights, Analytics and Enriched Fan
Experiences." IBM. July 6, 2017.
https://www.ibm.com/blogs/watson/2017/07/ib
m-watsons-ai-is-powering-wimbledon-highlights-
analytics-and-a-fan-experiences/

Smith, John R. "IBM Research Takes Watson to
Hollywood with the First Cognitive Movie Trailer."
IBM Think Blog. August 31, 2016.
https://www.ibm.com/blogs/think/2016/08/cogn
itive-movie-trailer/

"Solar Radiation Management." Wikipedia. Accessed July
13, 2018.
https://en.wikipedia.org/wiki/Solar_radiation_man
agement

"Sophia." Hanson Robotics. Accessed July 12, 2018. http://www.hansonrobotics.com/robot/sophia/

"The New Carbon Economy." CO2 Solutions. Accessed July 12, 2018. https://www.co2solutions.com/

"Tomra's Mineral and Ore Sorting Equipment for More Profit." Tomra. Accessed July 13, 2018. https://www.tomra.com/en/sorting/mining

"Trends in Atmospheric Carbon Dioxide." Earth System Research Laboratory. Accessed July 13, 2018. https://www.esrl.noaa.gov/gmd/ccgg/trends/gl_data.html

Walker, Jon. "Chatbot Comparison: Facebook, Microsoft, Amazon, and Google." Telemergence. March 29, 2018. https://www.techemergence.com/chatbot-comparison-facebook-microsoft-amazon-google/

World Economic Forum: https://www.weforum.org/

www.ingramcontent.com/pod-product-compliance
Lightning Source LLC
Chambersburg PA
CBHW022043190326
41520CB00008B/687